Evolutionary Critical Theory and Its Role in Public Affairs

Charles Frederick Abel and Arthur J. Sementelli

M.E.Sharpe
Armonk, New York
London, England

Copyright © 2004 by M.E. Sharpe, Inc.

All rights reserved. No part of this book may be reproduced in any form
without written permission from the publisher, M.E. Sharpe, Inc.,
80 Business Park Drive, Armonk, New York 10504.

Library of Congress Cataloging-in-Publication Data

Abel, Charles F.
    Evolutionary critical theory and its role in public affairs / Charles Frederick Abel and Arthur Jay Sementelli.
    p. cm.
    Includes bibliographical references and index.
    ISBN 0-7656-1046-9 (cloth: alk. paper) — ISBN 0-7656-1047-7 (pbk.: alk. paper)
    1. Public administration. 2. Critical theory. I. Sementelli, Arthur Jay, 1970– II. Title.

JF1351.A24 2004
351'.01—dc21
                                                                                 2003050598

Printed in the United States of America

The paper used in this publication meets the minimum requirements of
American National Standard for Information Sciences
Permanence of Paper for Printed Library Materials,
ANSI Z 39.48-1984.

BM (c) 10 9 8 7 6 5 4 3 2 1
BM (p) 10 9 8 7 6 5 4 3 2 1

Contents

Preface	vii
Acknowledgments	ix
1. Précis	3
2. Public Administration as Discipline and the Estrangement of Theory	10
3. Ontology and Theory in Public Administration	34
4. Critical Theory and Public Administration	64
5. Evolutionary Critical Theory	95
6. Evolutionary Critical Theory, Power, and Emancipation	121
7. Evolutionary Critical Theory and the "Good Society"	138
8. Evolutionary Critical Theory and Public Administration	160
Notes	175
Index	201
About the Authors	217

Preface

This book represents the culmination of an extensive collaboration between the two authors over the past five years. Throughout this process we have read, revised, and developed the ideas originating from a single paper at the Public Administration Theory Network. This in turn developed into a literature stream and ultimately into what is before you now. We owe a great debt to our colleagues at PAT-Net, as well as to Richard Box and others who encouraged us throughout the process.

At the beginning of our collaboration, after an extensive reading of critical theory, we discovered in our discussions that some of the elements we were exploring seemed to parallel the work of Thorstein Veblen. Initially, our impression was that he seemed to be an odd sort of fellow who did not quite fit in either as a sociologist or an economist, and that our discovery was a coincidence. Over time, we found that our initial discovery seemed to consistently fill a conceptual void in our readings of critical theory. It was at this point we decided to incorporate the ideas we learned through this "close reading" of Veblen and others into our understanding of critical theory. At the end of this process, we found that this incorporation evolved into a comprehensive supplement to the body of knowledge that is critical theory.

We then had to concern ourselves with the "so what" question. What did it matter for practical purposes that we developed an interesting take on critical theory? How could it help us broadly as a profession? During our writing process, we found that this "Evolutionary Critical Theory" as we coined it seemed to be exceptionally useful for addressing some of the classic arguments against critical theory. In addition, it seemed like this idea could help reconcile some major issues within public administration. However, this reconciliation might only happen if we could demonstrate first, that public administration is in fact a discipline and second, that it had ontological status.

With both a discipline and ontology, then public administration could meaningfully contribute to the body of knowledge as a profession and as theory generally. We found that by understanding of power and emancipation in this context, that public administration became essential to both theory and praxis due to its unique position in the world.

The discovery that public administration was essential to making a number of our arguments function in both theory and practice allowed us to deal with a few of the "big" questions people want to see in a book. We began to structure some alternative arguments for the identity, legitimacy, and epistemology of public administration. We hope the readers will find it useful, and suspect that a number will find our arguments both compelling and controversial. However, if our arguments at least get people to consider some of the possibilities of an Evolutionary Critical Theory, then we have achieved our goal.

Acknowledgments

We would like to thank the journal *Administrative Theory and Praxis* and the Public Administration Theory Network for supporting this book. *Administrative Theory and Praxis* graciously allowed us to build from articles that appeared in their journal to create several chapters in this book. In addition, the feedback we received from PAT-Net and its members allowed us to craft a much better piece than we could have otherwise.

Evolutionary Critical Theory and Its Role in Public Affairs

1

Précis

Administrative theory and practice are concerned with good governance. Now, many argue that theory in public administration is myth and illusion and that "public administration" itself is nothing other than a set of practices described and explained through a cluster of theories and methodologies that are borrowed from a variety of other social and natural sciences. As such, public administration constitutes a "field" and not a "discipline," and as a field it can claim no theory of its own. Instead, it must rely on theories developed by actual disciplines. Moreover, because what public administration ostensibly studies (i.e., administrative phenomena and behavior) is conventional in nature and not "brute data," it lacks any tangible reality beyond the conventional practices, behaviors, and discourses it studies. Because these conventions alone are the proper subjects of normative and empirical theory, public administration (a mere "artifact" assembled from them) has no ontological status and cannot itself be the subject of normative or empirical theory.

We will argue that this thesis is in error and that public administration is in fact a discipline developed to systematically work out and professionally apply the ways and means of good governance. Consequently, it can develop a theory of its own. We will argue further that its subject matter is ontologically constituted and that as a consequence its theory has ontological status. We will then argue that given the nature and status of its subject matter, public administration must employ synergistically and critically the methods of traditional social science, institutionalism, and hermeneutics to develop its theory and guide its practice. Finally, we will argue that a reconstructed form of critical theory ("Evolutionary Critical Theory") is especially helpful as a synergistic and self-critical agent and hence a powerful tool for administrative scholars and practitioners in their endeavors to realize good governance. To better present our argument, let us delineate briefly the key points that we will detail in the remaining chapters.

The complexity of "governance" is difficult to capture. Objectively, it is a dynamic process occurring between citizens and the structures of government (e.g., executive, legislative, and judicial organizations) and resulting in "a configuration of separable but interrelated elements—statutes, policy mandates, organizational, financial, and programmatic structures, administrative rules and guidelines, institutionalized rules and norms—which in combination establish the ends and means of governmental activity."[1] These elements take on different and changing configurations in different policy domains as the interests, roles, and priorities of the actors differ and change. As a result, the overall character of both the government and the society evolves.

"Good governance," however, is not quite so objective. While some still argue that good governance may be construed as conformance with a priori normative standards, most agree that what counts as good governance depends upon a "consensus attitude" toward governmental institutions and actors that is worked out by the citizenry through continued dialogue about competing attitudes, values, and beliefs. Hence, good governance is a matter of intersubjective agreement upon how to characterize the experience one has of governmental institutions and actors. This intersubjective experience, though fluid and cognitively less certain than positivist ontology might anticipate, is nevertheless ontologically constituted and therefore the subject of "first order" theory.[2] Hence, taking the intersubjective experience of good governance as its subject matter lends ontological status to empirical theory in public administration (it is thus theory) and provides an ontological grounding for normative critique. It is thus the standard that administrative theory and practice may hold up in order to judge institutional structure, process, practice, and outcomes, and the touchstone or reference point for the qualitatively different enquiries that must be made into how the intersubjective experience of good governance is secured and maintained.

Overall, then, we might say that "good governance" is whatever set of dynamic evolutionary processes so configures the means and ends of government as to secure and maintain the intersubjective experience of good governance. Hence scholars and practitioners of public administration must enquire into and address through theory and practice all behaviors, structures, processes, roles, norms, and values that impact significantly the evolution of that intersubjective experience.

Because good governance is determined intersubjectively, public administration must focus on society's institutions and networks, the locations of intersubjective exchange where "multiple social actors" negotiate and implement policy; deliver and receive social goods and services;[3] negotiate issues of representation, political control, and institutional legitimacy; and establish the means and ends of government.[4] And it must locate both its theoretical and practical endeavors at those points in institutions and networks where subjectivities (personal attitudes, values, and beliefs) meet, where the personal meets the social and institutional, and where the local (parochial) meets the broader society, as it is at these junctures that intersubjective experience is formed, deformed, and reformed.[5]

Overall, then, administrative theory and practice must take as its subject matter the evolution of the intersubjective experience of good governance and concern itself with all "loci of interrelationships" that affect that experience (e.g., in organizational, financial, and programmatic structures; administrative rules practices and procedures; institutionalized norms, statutes, policy mandates; the shared attitudes, values, beliefs, interests, and desires of those it affects both broadly and locally; and the shifting networks of interpersonal relationships within society). Since its founding as a field at the turn of the nineteenth century, public administration has in fact defined a distinct disciplinary matrix for itself that now encompasses this subject matter. However, because its subject matter requires it to accommodate qualitatively different objects within its study and practice (e.g., statutes and intersubjective experiences, finances, and beliefs), public administration must avail itself of qualitatively different methodologies that have developed over time as effective tools for describing and explaining these fundamentally different entities, their particular dynamics and their peculiar effects. Consequently, public administration must proceed methodologically by developing a synergy among institutional, behavioral, and hermeneutic approaches. It is insufficient for public administration to recognize the strengths and weaknesses of each and to employ each "where most appropriate," because its subject matter occurs at precisely the point where institutional structure, process, and power meet meaning, values, and behavior. Hence, employing these methodologies independently or only in traditional ways renders them less useful than does recognizing that while they were logically and culturally constructed for certain purposes, they may be used in different ways and "reconstructed" to some extent to serve other purposes.

In addition to the qualitatively different objects within its study and practice, public administration must contend with the fluidity of intersubjective experience. This has two implications for theory and practice. First, practitioners and theorists should "relax all preconceptions as to how the subject matter might behave, and permit themselves to be instructed by the subject matter" as revealed by the synergy among institutional, behavioral, and hermeneutic methodologies.[6] They must, in other words, "allow the subject matter to impress upon them its norms and to instill within them its categories."[7] Only after doing this should theorists "draw back to become objective and struggle to formulate a patterned explanation of the behavior that they have come to experience,"[8] and only then should the practitioners act. Second, both scholars and practitioners must remain cognizant of the fact that their understandings and choices of action are only provisional and contextually valid. Hence, theory must capture and practice must reflect the dynamic synergy among behavior, meaning, values, and institutional factors as they affect the flux and evolution of intersubjective experience and institutional change.

But as we have said, to systematically study the dynamic evolution of intersubjective experience, administrative scholars must synergize the methods of traditional social science, institutionalism, and hermeneutics. And to make sense of what would otherwise be a "blooming, buzzing confusion,"[9] practitioners need a theoretical framework. The problem is that by simply employing a methodology or a theoretical framework (no matter how carefully chosen), scholars and practitioners introduce certain preconceptions and value priorities into their endeavors that prepattern and so bias experience, explanation, and practice. To counter this inevitable bias, public administration's methodology, theory, and practice must be not only synergistic among institutional, behavioral, and hermeneutic approaches but self-critical as well.

"Critical Theory . . . is grounded in a perspective which has come to monopolize the categories of critique, praxis, and the fusion of theory and action."[10] This perspective addresses the intersubjective interplay of theory, culture, institutions, and society through interdisciplinary approaches (including the natural and traditional social sciences),[11] taken together (rather than read in contradistinction to one another). The theorists within this rich and robust tradition have employed the self-critical methods of negative dialectics, immanent critique, discourse analysis, praxis, and "transcendental pragmatics" to analyze the trends of Western civilization;[12] develop general models of social and political organization and process; develop general models of social change; develop models of how social, political, and psychological (subjective and intersubjective) dynamics work to impose both behavioral and ideological convergence (a social reality) upon groups and individuals with substantively different interests and needs; and suggest models of how the same processes might achieve convergence "rationally" (i.e., in a nonideological or nonimposed way).[13] Most important for our purposes, critical theory (again, taken as a whole) has engaged interdisciplinary approaches synergistically (negatively and positively), continually deconstructing and reconstructing them through critique in order to render them useful tools in concrete, evolving contexts. Hence, the employment of critical theory's self-critical methodologies by administrative theorists and practitioners promises to provide the synergy among institutional, behavioral, and hermeneutic approaches that is necessary to capture in theory and reflect in practice the dynamic synergy among behavior, meaning, values, and institutional factors affecting the flux and evolution of intersubjective experience and institutional change.

Unfortunately, there are several problems with employing traditional critical theory for these purposes. Perhaps the most significant is its nonevolutionary orientation. This orientation results from its consistent misconstruction (in our view) of both "power" and "freedom" (styled "emancipation") and the resulting tendency among critical theorists to overestimate (again, in our view) both the impact upon intersubjective experience of dominating cultural institutions and the desirability of a radically democratic society as the only "cure" under all circumstances. Thus, while critical theory implies the possibility of continual adaptive change arising from the intersubjective experiences of groups and individuals in concrete contexts, most (not all) of its theorists have consistently suggested only variations upon a single democratic theme as the solitary way to avoid the ineluctable "descent into barbarism" that will occur once all hope of agency is removed by the complete domination of a culture incarnating the Enlightenment's epistemology.[14]

These same misconstructions and tendencies account for critical theory's disdain for administrative institutions and its marginalization in the social sciences. Its consistently negative view of power and its individualistic view of emancipation, for example, leads it to conclude that administrative institutions necessarily mediate dominant ideologies, practices, and discourses, thereby keeping us from realizing a good society by perpetuating patterns of advantage and disadvantage. Hence, except in rare circumstances (e.g., religious institutions under certain conditions), most critical theorists neglect the role of administrative agencies in emancipating people from oppressive situations by, for example, coordinating their efforts to accomplish ends they could not accomplish alone. Regarding critical theory's marginalization, it has been noted empirically that institutional impacts are not as totalizing as critical theory suggests[15] and that a radically democratic solution assumes, contrary to the historical record, that (1) the intersubjective experiences of those "constrained" by a dominant culture predispose them to seek emancipation, (2) citizens will choose reasonable (i.e., nonoppressive) alternatives to currently oppressive institutions and habits of thought and action once their oppressive character is pointed out, and (3) "a better way" will be found through mutual agreement within the broadest possible constituency.[16]

There are other problems as well. Perhaps most important for administrative theory and practice is the fact that critical theory until Habermas and Honneth tacitly assumed the normative superiority of democratic governance under all circumstances and as a result enshrined certain radical forms of democracy as a teleological ideal. Both Honneth and Habermas recognized this problem, and both worked out conditions under which democratic forms may be normatively grounded. The problem is that in doing so each introduced certain "teleological moments"[17] into their versions of critical theory: Habermas by way of certain universalistic principles of discourse, and Honneth by way of certain universalistic principles of behavior. In both cases, the intrinsic desirability of some proffered ideal is held as the normative ground for democratic forms of governance. And this leaves us not significantly better off. Ultimately, Habermas recognized this fact, holding that his "universalistic principles" of discourse have "merely presumptive generality."[18] Consequently, they do not necessarily provide the required normative ground.

Equally important for administrative theory and practice are the questionable assumptions traditional critical theory makes about the mechanisms linking theory and practice. Many have pointed out, for example, that critical theory often overestimates both the ability and disposition of people to bridge the gap between norms and facts. The "lived experiences" of individuals are often manipulated to preclude the perception of such gaps, and simply providing insights into the inconsistencies between theory and practice in their lived experiences does not necessarily dispose or empower people to close them.[19]

Hence, while critical theory holds promise as the missing link in public administration's methodology, a link that might render theory and practice both self-critical and synergistic among institutional, behavioral, and hermeneutic approaches as well, its traditional form falls just short of the mark. However, its promise and its undeniably useful insights into the character of Western philosophy, political theory, social theory, cultural theory, and the search for an interdisciplinary social science suggest that it might be reconstructed without the baggage of its radically democratic teleology, its supposition that administrative institutions and power are oppressive necessarily, and its overly individualistic understanding of freedom. Absent this baggage, we might expect that a reconstructed critical theory will better describe and explain the historical record, close the theory-practice gap, and prove useful to public administration as a synergistic and self-critical agent.

Toward such a reconstruction, it is important to note that the writings of Thorstein Veblen share the ultimate goals and concerns of traditional critical theory (e.g., a society that recognizes the equal dignity, worth, and honor of individual human beings by being continuously responsive to the real and changing needs and interests of its people) and employ many of its methods (e.g., the synergistic use of immanent critique and negative dialectics) sans its teleology, its disdain for all administrative institutions, its thoroughly negative view of power, and its individualistic concept of freedom. Veblen's writings suggest instead that what counts as good governance, at any time and in any place, evolves "endogenously" (as do needs and interests) from the intersubjective experience of those in context. Consequently, it may take many forms (some that may be perhaps less than radically democratic), it may involve the positive use of power and institutions, and it may realize many different modes of freedom. From this basis, then, it seems we should be able to reconstruct critical theory into an agent driving the sort of synergistic and self-critical process required methodologically for good theory in public administration.

But such a reconstruction promises still more. The development of an Evolutionary Critical Theory for public administration affords us opportunities to address a number of the discipline's central questions. For example, by delineating public administration's subject matter and confirming its ontological constitution, Evolutionary Critical Theory resolves public administration's ostensible identity crisis. In addition, by identifying public administration's subject matter as the intersubjective experience of good governance, we address the big question—its legitimacy or proper place in governance. Since public administration's role is to foster the intersubjective experience of good governance, the discipline is responsible to society and the public *as a whole,* rather than to some particular political construct such as Congress or the presidency. Finally, this reconstruction promises to put the argument between empiricists and theorists about the extent to which one can usefully inform the other to rest. The argument, of course, grows from the belief among some scholars that "theorizing" is of limited use[20] and the belief among others that "brute" empiricism is meaningless without a theoretical framework.[21] This argument has been particularly bothersome since it has allowed for a number of public administration's critics to challenge its legitimacy, identity, and disciplinary status.[22] However, these challenges are resolved and the differences are overcome by the synergistic use of theory and practice accomplished through Evolutionary Critical Theory.

To summarize, it is our thesis that public administration would gain theoretically and practically by *synergistically* employing the methodologies of traditional social science, institutionalism, and hermeneutics, and that such a synergy could be accomplished through the use of a reconstructed critical theory as synergistic agent. Through a succession of chapters we will seek to substantiate this thesis, beginning with an explication of (1) public administration as a discipline, (2) the ontological nature of its subject matter, and (3) why the theory and practice of the discipline might be advanced, and the subject matter better understood, through a synergistic use of methodologies. With this foundation in place, we will proceed to explain (1) the promise and limitations of critical theory as a synergistic agent (given the nature of public administration's subject matter and the locus of its endeavors), (2) how critical theory might be reconstructed as Evolutionary Critical Theory, and (3) how Evolutionary Critical Theory overcomes the above-mentioned limitations of critical theory and is thus more useful to the discipline. We will then conclude with some suggestions regarding how employing Evolutionary Critical Theory as a synergistic agent among methodologies affords us the opportunity to address a number of the discipline's core theoretical and practical concerns.

2

Public Administration as Discipline and the Estrangement of Theory

Theory and Public Administration

Theorists are estranged in public administration. They are estranged because most practitioners and much of academia are convinced that "theory" is dead. It is dead, the argument goes, because the culture, discourse, and interest-bound nature of theory precludes anything more than situationally[1] derived "islands of theory,"[2] or time- and context-bound "middle range theories"[3] that are "intermediate to the minor working hypotheses evolved in abundance during day-to-day routines . . . and the all-inclusive speculations comprising a master conceptual scheme."[4] This situation arises and the argument continues because public administration is typically on the horns of a dilemma. On the one hand, public administration seeks to ground its knowledge in observation and experience; hence, it rejects formalism, metaphysics, and all other speculative a priori claims. On the other hand, it seems clear that facts do not present themselves unequivocally and that understanding what we are observing or experiencing requires just the sort of abstraction, generalization, and intersubjective agreement afforded by the very synthetic apriorities that public administration rejects. The only way to resolve this dilemma, the argument concludes, is to characterize theories as cognitive tools (heuristic devices and cognitive lenses) employed toward coping with the complex chaos we individually experience from moment to moment. As heuristic devices and cognitive lenses rather than global explanations or systematic statements of underlying principles, theories become useful in the discovery, organization, and explanation of what we individually observe, and they provide the intersubjective framework necessary to describe, relate, and generalize about observations and experiences among individual "selves."

Of course, being tools and devices, theories do not necessarily accord with "reality" (except, perhaps, by happy accident), and they are thus neither right nor wrong. Nevertheless, the more useful a theory's analytical concepts and models prove for immediate purposes, the more they tend to reify into the "correct" or "rational" way of seeing and doing. By thus inveigling us into fitting our experiences and observations to their concepts and models, theories encourage us to forget that what we observe and experience is both context-bound and dependent to a significant extent upon the values, interests, needs, and desires of the observer (conditioned, of course, by culture and extant institutions). Hence, theories not only make sense of reality, but simultaneously distort it as well. In this way, then, theories quit being neither right nor wrong and become both right and wrong.

This makes a significant difference. As abstractions reified into correct and rational ways of seeing and doing, theories exert a strong conservative tendency. They tend to reinforce the social, political, and economic status quo by persuading us to accept their reifications as given. However, as the *quo* never itself reifies, the *status* is always to some extent estranged from what is experienced and observed. To resolve this problem, it is thought that theory must include the subjective "lifeworlds" of people in their particular social, economic, and political contexts and that it must reconstruct the particular conventional contexts, concepts, and models that inform and express those "lifeworlds." Hence, because theories are heuristic devices and cognitive lenses, and because they must comprise the intensions, meanings, and conventional contexts of their subjects, theory is necessary limited in descriptive and explanatory scope to the particular context and actors about whom it is concerned. And as both change regularly and vary from place to place, from role to role and from discourse to discourse, theories proliferate.

In this manner, then, are theory and public administration estranged. So far as public administration is concerned, both the Enlightenment hope of applying theory to practice and the Hegelian vision of theory merged with practice are dashed by the antithetical view that there is a "natural and unbridgeable" theory-practice gap.[5] The most that can be hoped is that scholars and practitioners will strike upon or invent tools useful for the situation and the moment and that they will come to recognize when their tools no longer serve and should be discarded. This idea of theory as a discardable tool is bolstered by postmodern doubts about the ability of science, philosophy, religion, or any other form of reason, society, civilization, or culture to address impartially any really profound question. As a consequence, recourse to theory for sensible explanations and reasonable justifications for including any particular content, or articulating any particular issue, or accepting any particular answer as theoretically and practically acceptable in any particular field, is displaced by recourse to the sheer custom, habit, interest, and "discourse" of particular groupings of individuals.[6] Of course, as we cannot specify any particular content of pubic administration, and as we cannot distinguish a definite set of issues that the field must address other than those historically and culturally given by custom, habit, and discourse, theory and theorists are marginalized, alienated, and estranged from the scholarship and practice of the field.

Given the nature of theory (as theory), then, there are necessarily at any given moment many different, sometimes competing, definitions of and theories about public administration's subject matter; and as these different theories and definitions necessarily remain relatively autonomous, public administration is always experiencing an "identity crisis" (e.g., Marini; Box; White, Adams, and Forrester).[7] Some even suggest that the ongoing confusion engendered by this crisis *is* public administration's identity.[8] Be that as it may, this lack of identity is said to preclude the development of either a disciplinary identity or a theory of public administration. Instead, the field is said to encompass a vast and heterogeneous range of subject matters and an equally diverse array of paradigms, "schools of thought," theories, "narratives," and "guiding ideas," many of which are considered artifacts of questionable ontological status, and all of which are clustered loosely together, though they lack necessarily any common set of concepts, practices, purposes, values, or methodologies upon which scholars and practitioners might settle. In brief, because the content of public administration is unable to be specified, public administration lacks an identity and therefore fails to qualify as a discipline; and because it is not a discipline, any theory about public administration is impossible.

We will argue that this thesis is in error and that public administration is actually a coherent discipline defined by a lucid "disciplinary matrix." We will argue further that, consequently, although this matrix is historically and culturally embedded (and despite the consequent lack of timeless, philosophically accessible transcendental foundations), there can be meaningful theory of a particular sort in public administration. Specifically, we will argue that a coherent and evolving disciplinary matrix constitutes the identity of public administration, that the subject of this disciplinary matrix is the intersubjective experience of good government and how that experience might be accomplished, and that this subject matter has ontological status. Because public administration's subject matter has a distinguishable ontological identity, its evolving disciplinary matrix provides the structure necessary both to maintain continuity among differing paradigms, "narratives," and "islands of theory" and to provide a basis for evaluating their meaningfulness and commensurability. As meaningful theoretical commensurability thus resides *in* the evolution of public administration as a discipline, a theory is possible;

and given the ontology of public administration's subject matter, that theory is necessarily critical and evolutionary in nature. We will then suggest that such a theory ("Evolutionary Critical Theory"), grounded in the thinking of Thorstein Veblen and capable of closing the theory-practice gap, is most conducive to public administration's identity.

The Argument in a Nutshell

As a first step in substantiating this claim, then, we will argue that the view of public administration as a loosely related "cluster" is largely a reaction to what is variously styled the "orthodoxy," the "bureaucratic paradigm," the "conventional view," and the "traditional model" of public administration, and that this orthodoxy was developed and continues to serve very real needs and interests. Specifically, we will argue that the orthodoxy is in the nature of a "form" (a la Plato) that both scholars and practitioners developed and continue to employ as a touchstone, heuristic, and signifier, of a much more sophisticated and evolving disciplinary matrix that is both operative in practice and definitive of the field as a distinct discipline. We will then show analytically and empirically that the disciplinary matrix we perceive is organized coherently around both a particular subject matter and a core complex of purposes, values, practices, and behaviors and that this core grounds a periphery of related complexes identifiable by their family resemblances to the core. In the process, will demonstrate how both the core and the periphery are negotiable and regularly challenged and how all attempted "radical renegotiations" of the core have failed while much successful renegotiation (evolution) occurs regularly at the periphery. As a consequence, the clustering of approaches in reaction to the orthodoxy is not only unnecessary and confusing but obfuscates the disciplinary matrix defining public administration to a far greater extent than does the orthodoxy itself. More important for our purposes, the fact that public administration is in fact a coherent discipline implies the possibility of its developing its own distinguishable theory.

Rotating Plato: The Orthodoxy and Its Uses

Plato introduced the notion of a "form" as an ideal way of being that exists beyond the material world but gives things in the material world their particular nature or character. The things of the material world are always in continuous approximation to the form but never realize it in the physical realm. For pragmatic reasons (political, social, and academic), scholars and practitioners of public administration rotated Plato by constructing an ideal form behind the reality of administrative practice. The features of this form may be succinctly stated. Politics and administration are distinct. Politics is about developing public policy and public administration is about executing it in a successful way (i.e., so as to engender the intersubjective experience of good governance). Public administration is thus a tool at the disposal of those elected to exercise power, and it is therefore proper for political institutions and processes to set policy and control administration. Nevertheless, to ensure the accurate, efficient, fair execution of policy (i.e., to engender the intersubjective experience of good governance), politics should not be part of the administrative process itself. Rather, the processes of policy execution should be studied "scientifically" in order to derive the most rational, economic, efficient means of administering policy.

The form persists and is always under assault. Most notable are the well-known critiques of Dahl[9] and Simon[10] as embellished by Waldo.[11] The form is understood, in other words, to bear a poor resemblance to reality. Nevertheless, it persists; and for our purposes, the important thing to note is that the orthodoxy persists as a highly simplified version of a Kuhnian disciplinary matrix. As Kuhn elucidated the various uses of "paradigm" within his original theory, he explained that there are "paradigms" in the narrow sense and "paradigms" in the fullest sense.[12] Paradigms in the narrow sense are best thought of as only one element in the definition of a discipline. Disciplines are particular communities united by a disciplinary matrix (a paradigm in the fullest sense). This disciplinary matrix includes symbolic generalizations (formal or readily formalizable components), models (deliberate simplifications of reality), and values and exemplars (paradigms in the narrow sense or concrete problems that can be solved by the methodological forms provided by the community).[13] The orthodoxy is itself an obvious simplification of reality expressing a set of political, methodological, and institutional values that may (perhaps too facilely) be formalized and researched according to well-understood (though, as we understand, often inappropriate "scientific") exemplars.

We believe that this oversimplified version of public administration's disciplinary matrix was developed and persists (consciously and unconsciously) both because it is a useful signifier of the discipline's more complex and realistic disciplinary matrix and because it serves very real interests as a touchstone and heuristic. First, it is a useful rule of thumb for marking off the boundaries of public administration as a field of study and for promoting an independent relationship between elected officials and administrators in practice. Montjoy and Watson, for example, argue that the orthodoxy provides a rationale for insulating administrative practice from politics and "remains important as a normative standard in the profession of local government management."[14] Next, as Garvey points out,

"Public Administration" involves necessarily a certain unique dilemma. Administrative action in any political system, but especially in a democracy, must somehow realize two objectives simultaneously. It is necessary to construct and maintain administrative capacity, and it is equally necessary to control it in order to ensure the responsiveness of the public bureaucracy to higher authority.[15]

The orthodoxy sets quick and dirty behavioral norms for administrative practice that orient practitioners toward the practical resolution of this dilemma. Third, the orthodoxy is a convenient starting point for the study of what public administration actually is and does. It provides both a simple introduction to public administration's epistemological and normative orientations and also a simple model of public administration's role in both the state and society; it is a model that may be thrown away without harm as deeper understandings emerge. Finally, the orthodoxy provides citizens with an efficient way of organizing and simplifying their understanding of bureaucracies and how they work in order to come to an intersubjective evaluation of the governance they are experiencing. The orthodoxy is efficient both because it requires relatively little information to understand and because it enables ordinary citizens to make good political judgments about bureaucracies and their behavior even when they lack general political acumen or encyclopedic information about the specific bureaucracy in question. The orthodoxy's simplified disciplinary matrix, in brief, is an adaptive and helpful symbolic and discursive response to a complex reality.

Public Administration as Nothing in Particular: The "Clustering" Approach

As useful as the orthodoxy is, however, it is insufficient as anything other than a signifier or starting point. Scholars and practitioners require a deeper, broader, and more complex understanding. To satisfy this requirement, many scholars and practitioners of public administration cluster whole complexes of theories, paradigms, values, and practices under the term. Waldo, for example, argued that the scope and nature of public administration as embodied in its pre–World War II orthodoxy were dubious.[16] He then went on to point out that in a postwar reaction to the orthodoxy, public administration went beyond political science for its theoretical, prescriptive, and intellectual base, delving into social psychology, economics, sociology, and business administration.[17] Ostrom argued that public administration faced a crisis of proliferating theories, methodologies, and epistemologies.[18] Earlier, Golembiewski had suggested that public administration develop a "family of mini paradigms" to encompass its work and study.[19] Henderson had suggested that the heterogeneous literature might be made sense of in terms of three emphases (structural, behavioral/environmental and organizational), although he added, "It is not easy to tie together a vast number of publications whose chief similarity is dissatisfaction with pre-war scholarship" (i.e., the orthodoxy).[20] Not trying to tie things together but simply to get things organized, Rosenbloom, concluded that there were three competing approaches to public administration—political, legal, and managerial—each with its own values, methods, and intellectual tradition.[21]

The postmodern turn in public administration adds to the confusion and further elucidates the poverty of "clustering." Postmodernists oppose the idea that language can refer clearly to any definite entity. All definitions are socially constructed, deconstructed, and reconstructed on a continuing basis, and variation in the meaning or the extension of a concept is a function of alternative evaluative perspectives. To understand the historically and culturally constructed meaning of something, deconstruction is employed, not to elucidate in the sense of attempting to grasp a unifying content or theme, but to elucidate cultural biases, oppressive power relationships, and dominating epistemologies that secure patterns of advantage and disadvantage. Ultimately, these biases, relationships, epistemologies, and practices are considered the definition of public administration.

For our purposes, the major problem with postmodernist and all other clustering approaches is that they permit neither comprehension nor the development of an intersubjective experience of good governance. If, as these scholars suggest, we can construct public administration only by clustering a heterogeneous population of events under the term according to our value orientations, then disputes over the proper use or extension of the term must be value disputes and both the meaning and the identity of public administration are endlessly disputable. An untidy package emerges. First, public administration extends to an array of heterogeneous practices clustered for politically or culturally biased ends. Consequently, including or excluding any practice from public administration is a function of alternative value-bound perspectives. Thus, public administration is doomed to having multiple, incommensurable meanings (identities) covering whatever cluster of practices we might value. The problem, of course, is that this renders the identity of public administration incomprehensible. It is everything, and so it is nothing.

For example, consider what we are to comprehend as "public." The practice of generalizing our behaviors into "public" and "private" categories is so woven into our culture that it seems natural in the sense of ontologically fundamental. Postmodernists ask us to deconstruct these generalizations, suggesting that if we consider other mind-sets (other culturally derived ideas of the way things are), we might remove certain distortions and delusions that the generalizations introduce. As Farmer puts it,

> A clearing away of delusions in public administration thinking is an important benefit . . . promised by deconstruction and the postmodern turn. Liberating public administration thinking from entrapment in the efficiency metaphor, with the binary opposition of efficient-inefficient, is [one] example of this kind of benefit . . . another is release from the entrapment of public administration thinking in . . . such binary oppositions as productive-unproductive and private-public . . . another example is the liberating of public administration thinking from its entrapment in . . . such binary oppositions as autonomy-heteronomy. . . . Yet another example (suggested to me by Charles Fox and Hugh Miller) is liberation from the entrapment of public administration thinking in the narratives of money, with such oppositions as solvency-insolvency and funded-unfunded.[22]

According to the "postmodern turn," then, it is "distorting" to construct the term public administration to include not only any public/private generalizations, but any distinct element of efficiency (or inefficiency), productivity (or nonproductivity), autonomy (or reliance), solvency (or insolvency) as well. Rather, it must be deconstructed to include all of these things and everything in between. In brief, the term must refer to nothing in particular in order to avoid "distortion" and "delusion."

Relieved of these delusions, however, we have no idea what to do, study, or evaluate. Our choices are infinite, but we have no way of exercising any choice at all. Worse, any choices we do make will either be arbitrary or value-biased and therefore distorting and deluding. Still worse, the very act of thinking as "others" think must also be distorting as it singles out one way over others and requires that we value some particular other way of thinking over both our own culturally constructed way of thinking and all other ways as well. Worse, worse, worse, it must necessarily delude one into thinking that perhaps more harm than good comes from our culturally woven construct, but this cannot be the case as all culturally based ways of thinking are distorting and deluding. Hence we are caught in an endless spiral of undecidability. We simply cannot choose among, say, public administration as something collectivist, or something anarchistic, or as something communitarian. We must simply float among all possibilities. In fact, public administration becomes so undecipherable as to include its negation:

> Beyond the clearing of these and other delusions, deconstruction raises the prospect of antiadministration [sic]. Deconstruction, as an element in a postmodern attitude, suggests the possibility of a juxtaposition of "administration and antiadministration" that can provide significant change in public administration practice. The metaphor antiadministration is suggestive of antimatter or anti-particles . . . [and can be understood as] administration that is simultaneously directed at negating administrative-bureaucratic power and that negates the rational-hierarchical Weberian outlook.[23]

In this regard, it is interesting to note that physicists tell us that should we come into contact with antimatter we will disappear. As this passage indicates, so does any decipherable meaning of public administration. As Farmer admits, "The upshot of deconstruction is radical indeterminacy."[24]

Thus, despite the efforts at containment by writers like Henderson and Rosenbloom, the postmodernist turn and the clustering approach lead many to conclude that the scope of public administration seems unlimited.[25] At best, we might say along with White that public administration consists largely of local narratives produced by a multiplicity of methods and perspectives. What is missing is some framework, ground, or core interconnecting the local narratives and multiple methods and preserving "a greater sense of public administration as a whole."[26] However, it is worth contemplating the possibility that in the rush to replace the orthodoxy with alternative complexes of purposes, practices, values, and behaviors, insufficient attention was paid to what the orthodoxy signified. It is worth considering, in other words, the possibility that the orthodoxy is a sign and not a theory, a model, or a definition; and that this sign points to both a ground or core and a framework interconnecting concrete narratives, practices, and academic studies.

Public Administration as Loosely Related Phenomena: The "Family Resemblances" Approach

Most promising in revealing these possible interconnections is the "family resemblance" approach. This approach was developed by Wittgenstein to resolve certain philosophical conundrums.[27] These conundrums, as it turns out, have their origin in the notion that things described by a common word must have a common essence. As Wittgenstein noted, this notion is not supported by ordinary language. Directing our attention to the word "game," for example, Wittgenstein pointed out that "if you look at them [games] you will not see something that is common to all, but similarities, relationships, and a whole series of them at that."[28] Concepts thus develop "as in spinning a thread we twist fibre on fibre. And the strength of the thread does not reside in the fact that some one fibre runs through its whole length, but in the overlapping of many fibres."[29] So a common word need not denote an essence but only a grouping or twining of resemblances among the uses of the term.

Wittgenstein further suggested that the particular way that any given concept becomes twined together is determined neither externally (say, by the nature of the world), nor individually (say, by a person's values). Rather, it is determined in concrete social practice by both the ways a word is used and the ways that others respond. That is, it is "only in the stream of thought and life [that] words have meaning,"[30] and using words in particular ways is thus part of a particular "form of life."[31] Put another way, the meaning of a word "explains the use of the word" in the language games embedded in a "form of life's" concrete customs and practices.[32]

As customs and practices "drift," the uses of words are not "fixed" in the strict sense of the word. Novel uses may be presented and "negotiated into" the language game, but to be successfully "negotiated in" (and hence to constitute a "legitimate" use of the term), all such usages must be intelligible; and intelligibility rests upon "family resemblances" to the ways the term has been previously used. Thus, although there is not any single "form" (in the Platonic sense) that legitimates each specific use of a word, there is nevertheless a "genealogical" relationship linking each intelligible use with all other current and previous uses. In brief, although the scope of any word is undetermined, all uses (to be intelligible) must be linked through connections which may fade, but which still provide a distinguishable trail.[33]

Under this approach, what counts as public administration is socially learned and negotiated within a form of life just as what counts as a game is learned and negotiated. Disparate complexes of practices, behaviors, and purposes that strike us as similar but not the same motivate us to include them together as public administration, despite their lacking any definitive crux. An original or settled set of complexes may be socially negotiated (consciously and unconsciously) over time and be generally accepted as the definition. On the one hand, imagining that this definition necessarily includes all of the possible complexes (i.e., that public administration has a definite scope) or that any one (or any part) of the complexes is needed (i.e., that there is an archetypical or paradigmatic or essential complex) is wrongheaded; any complex may be removed without damage to the definition, as no single complex is necessary or determinative of the term's use. On the other hand, certain complexes are culturally (not ontologically) fundamental. That is, to make any sense to people discoursing about public administration, the use of the term may not be arbitrary (i.e., include just anything based, say, on our value orientations). It must be socially and culturally coherent. It is not essentially any particular thing, but it can not be just anything at all. Any transformation in the use of the term public administration, then, is (intersubjectively) coherent exactly to the extent that it bears reasonable family resemblances to other current and previous uses of the term within our form of life.

As a term subsuming complexes related in a family way, then, public administration may constitute a fluctuating set of practices, activities, pursuits, and purposes related in a family way but sharing no essential core. Hence, public administration may be describable as a form of power and authority, a bargaining transaction among equals, a set of hierarchical management transactions, or a democratic set of organizational change practices (though no common property seems to extend through all of them), so long as the description of each bears distinguishable family resemblances (no matter how faded) to the ways that the term public administration has been used before. Employing the term without such resemblances would be so distorting as to mislead, confuse, and obfuscate. To say intelligibly that "public" includes "private," for example, requires some explanation of how the private resembles the public or has public characteristics or dimensions.

A significant body of literature argues for this approach to both defining and redefining public administration. Farmer, for example, makes explicit use of this approach in moving toward a profound redefinition of public administration.[34] As he says, "a radical change is needed in the way we conceptualize the role and nature of public administration theory."[35] He seeks to accomplish this by pointing out that public administration is a "form of life" sporting its own "language" (broadly defined to include not just words and grammar but "styles"). He grounds this approach solidly in Wittgenstein's notion that language is embedded as a constituent of any "form of life" and suggests that in order to change public administration we must alter the language game we play. He argues that this move is promising given that "language is not a closed system" and that there is always some latitude available in constructing "any connection between . . . what we mean by the signifier PA and what is signified by the term PA."[36] By exploiting this latitude through what he calls "the art [of] reflexive interpretation,"[37] Farmer seeks to elucidate other ways of "seeing" (defining) public administration and to begin negotiating their acceptance as definitions of the term.

Similarly, Fox and Miller, concerned about how we might make knowledge claims in public administration, point out that claims are made through linguistic discourse and that the "context of daily life contains habit and routine . . . [or] recursive practices . . . that generate linguistic customs that constitute participants' meaning making."[38] Hence, the stable background of recursive practices, including recursive ways of employing public administration in discourse, defines not only the meaning of any knowledge claim we might make within "public administration," but also the meaning of the term itself. Of course, the "background of language, customs, and practices ... can ... be changed."[39] However, "claims make sense only when they fit into some preexisting conceptual scheme taken as coherent by an epistemic community";[40] and given the "disorientation that comes from changing everything at once, we can renegotiate only some small part of the totality of habits and norms we have developed over time."[41] Hence, the meaning of public administration is neither a "representation of reality" nor an arbitrary utterance. Rather, its meaning is "tethered to the rules of situated language games and to the human participants" and cannot be understood "outside of participants' meaning making in a community."[42]

Convinced that by this "discourse theory" of public administration Fox and Miller offer a viable alternative to the orthodoxy, Hansen is interested in how one knows "that authentic discourse [and hence a "public administration" discursively defined] is being practiced."[43] In brief, how does one define "discourse" and hence redefine "public administration" discursively, so that we know it when we see it? Fox and Miller suggest that we use certain "warrants" for the proper use of the term. Specifically, they argue that we have true discourse (and hence a truly discursive definition of "public administration") when the participants are sincere, intend to make relevant contributions, pay willing attention, and have the ability to make substantive contributions.[44] Concerned that these warrants may have a reliability problem (e.g., how do we know someone is being sincere?), Hansen suggests that we require more and different kinds of reference points that might be observed and interpreted through their "organic qualities and place value ... [in] practical usage."[45]

For our purposes, what is important is that Fox, Miller, and Hansen all ground a proposed redefinition of "public administration" in intersubjective referents both to uses of the term in actual practice and to the intersubjective meaning of the term as simplified in the orthodoxy. That is, they are suggesting a redefinition of "public administration" bearing sufficient family resemblances to certain features of the orthodoxy so as to suggest that they are talking about the same thing from a reasonably different (perhaps better) intersubjective perspective. This is achieved by proposing renegotiations of, for example, what counts as "administering policy" and what counts as administering it well. They are also renegotiating what counts as the political processes that should "control administration" and how we should study the policy and administrative processes in order to improve administration. Thus, we can perceive both the novelty and distinctness of the proposed redefinition (it is its own person, so to speak) and the features of the orthodoxy that gave it birth. Consequently, we are not confused, and the meaning of "public administration" is not obscured. We understand that we are talking about the same thing; the only question is whether we will intersubjectively accept this new person as a relative.

This approach, then, holds great promise. As Wittgenstein suggests, explaining the meaning of "public administration" should involve "explaining the use of the word" in the language games that are embedded in the concrete customs and practices constituting our "form of life."[46] If we cannot do this, "public administration" is up for grabs and we may all cluster away until something settles out or we weary of the effort. However, if we can make such an explanation, "public administration" has an intelligible meaning, and novel uses of the term must bear sufficient family resemblances to its trackable uses if they are to make any sense. Moreover, not only would "public administration" have an intelligible meaning (and hence a definition), but we would also have at our disposal helpful insights into both how such a meaning arose and how that meaning may continue to change coherently. That is, tracking the family resemblances back in time may lead us to what the orthodoxy signifies (unless, of course, the orthodoxy was in fact the origin). Thus public administration would gain a ground, a definitive (though negotiable) core, and a scope that not only precludes inroads by other fields into public administration's domain but also provides answers (positive and negative) to critics attempting to either "negotiate in" certain extensions of the term or capture the term for their own purposes. Better still, conceptualizing public administration in this way would remove the distinctions among theory, practice, and analysis by removing the distinction among meaning, language, and action. This, in turn, would allow for a definition of public administration that is both a theory and an empirical research paradigm.

Some Thoughts on Using Family Resemblances in Defining "Public Administration"

So far, then, it is reasonable to conclude that in order to have a ground and a coherent core, scope, and research paradigm, the term "public administration" must have a history of uses in concrete contexts that is trackable through family resemblances to something other than the orthodoxy (i.e., the orthodoxy must be a signifier and not the origin of the term's use). Does "public administration" have such a history of trackable uses? That is, does it in fact have a meaning, or are we still at square one? If it has a trail of uses, how does this trail affect current negotiations about how the term should be employed intelligibly?

In answering these questions, it is interesting to ponder what counts as family resemblances. In Wittgenstein's account, what counts as family resemblances seems to be socially learned and negotiated much as what counts as a game or the color red is. Disparate complexes of practices, behaviors, and purposes that strike us as similar but not the same motivate us to include them together under a term despite their lacking any definitive crux. An original or settled set of complexes may be socially negotiated (consciously and unconsciously) over time and be generally accepted as the definition. However, those who imagine that the definition necessarily includes all of the complexes (i.e., that public administration has a necessary scope) or that any one (or any part) of the complexes is needed (i.e., that there is an archetypical or paradigmatic or essential complex) are wrongheaded; any complex may be removed without damage to the definition, as no single complex is necessary or determinative of the term's use.

Given all this, it is interesting to note that public administration departments and programs currently exist at many universities and that these universities accredit undergraduate and graduate public administration courses. It is also interesting to note that a number of local, state, regional, and national associations hold conferences and publish journals in the field. Moreover, some of these associations certify public administration programs and regulate both such certification and the practice of public administration. Finally, it is interesting to note that none of these programs, courses, associations, or journals takes the orthodoxy seriously as a disciplinary matrix. All employ it as a foil, heuristic, or touchstone, taking understandable delight in both elucidating its shortcomings and employing it effectively as a pedagogic tool. Thus, it is difficult to imagine that some complex or group of complexes beyond the orthodoxy did not at one time, at least, constitute the common origin of the current set of complexes we call public administration, and that such an origin may not be discovered by tracing the genealogy of the term. As Wittgenstein puts it, for example, new streets may be added to a city's core, thus redefining the city, and even the original streets may be restructured, obliterating the city's original definition. Nevertheless, the original definition is the foundation for the definition as it now exists, and it exerts an impact even once obliterated.

In brief, given the uses of the orthodoxy and the fact that there is an ongoing practice called public administration, it is reasonable to suspect that the various intelligible uses of "public administration" have a common origin functioning as a core concept. While there may be no ontological essence to this core (though we will argue that there is), there may still be a cultural essence, a common source or a single linguistic root from which the current definition is thrown up. Such a root might not only illuminate the meaning of the term but legitimate the inclusion of other complexes (connect them together) into the definition as well. Just as the family resemblances among a group of people is more understandable if the family's roots are discovered, so are those among the complexes of practices, behaviors, and purposes we call public administration. Such origins may, of course, be left behind at some point, but their exclusion through renegotiation of the term's use would seem to require an especially strong justification. Consequently, they may seem more central to the use of "public administration" than other complexes employed in or claimed as part of the term's definition, and radically different complexes would tend to obscure and confuse rather than elucidate the meaning of the term.

Similarly, it is difficult to imagine that of the complexes that are less central, some are not more important than others. No one complex or group of complexes may be necessary, but it seems that some are more important than others for successful reference under ordinary circumstances. Additionally, certain complexes might be important to maintaining certain propositions about public administration that are significant for pragmatic, political, or philosophical reasons. Certainly, some set is necessary for benchmarking if only as a point of departure for a complete reworking of the definition. There seems, in other words, sufficient reason to suspect that there is a core to the set of complexes whose retention is required for the maintenance, extension, shifting, or renegotiation of the definition. Such core complexes might, therefore, be said to have a relatively more fixed status for all ordinary uses of the concept simply because the absence of these complexes threatens to render the remaining complexes unintelligible or at least inadequate as successful referents. Put another way, removing the core complex(es) renders the term unusable or indistinguishable, while removing less central complexes leaves the referent both distinguishable and useful.

If this line of thinking is correct, it seems reasonable to conclude that there are identifiable, central, important complexes of practices, behaviors, and purposes to the concept of public administration, even though there is no essence to the term and even though the set of complexes is porous. Disagreements over what is properly public administration can therefore be sorted into three categories. First, disputes over complexes that are not at the core are nonessential disagreements. While they may seek to extend or restrict the uses of the term, they do not alter the usefulness or distinguishability of the concept. Disputes over core complexes, on the other hand, are essential. They require complete conceptual or paradigmatic shifts along the lines articulated by Kuhn.[47] Finally, confrontations of the core by complexes bearing little or no relationship (i.e., insufficient family resemblances) to the core (e.g., impositions of complexes from other fields or disciplines, attempts to capture the term in order to promote diametrically opposed values) constitute not disputes within public administration but attempts to subjugate the term to radically different ends, if not to obliterate it altogether and replace it with something completely different.

The Common Origin and Referential Necessities: The Intersubjective Experience of Good Governance

While public administration has historical antecedents from which it was finally thrown up, it is well documented that, "when it comes to the development of an independent administrative discourse, the Americans take the lead around 1900."[48] Goodnow and Wilson are generally considered the "founding fathers,"[49] and during this period public administration conceived of itself as "the management of men and materials in the accomplishment of the purposes of the state."[50] In concert with this idea, public administration took its identity from a "cluster of related tenets" that were actually somewhat different than those embodied in the orthodoxy.[51] Briefly, public administration thought of itself as an endeavor distinct from the activities of policy-making institutions but nevertheless involved in all things political. As Wilson explained, "Public administration is detailed and systematic execution of public law . . . [though] the general laws . . . are obviously outside of and above administration. The broad plans of governmental action are not administrative; the detailed execution of such plans is administrative."[52] However, while Wilson thus sought to distinguish public administration from the politics of lawmaking, he also felt that "the real function of administration is not merely ministerial, but adaptive, guiding, discretionary. . . . It must accommodate and realize the law in practice."[53] Consequently, to fulfill its proper function, public administration required "large powers and unhampered discretion," and politicians, legislatures, and even courts "should not be suffered to manipulate its offices."[54] Thus, in exercising the power necessary to put law into practice, public administration necessarily accommodates, guides, and adapts the law. It is thus intimately involved in the politics of policy making through practice.

Goodnow echoed the view that administration involved politics. Though he is often credited with first making a distinction between the two, Waldo,[55] Golembiewski,[56] and Caiden[57] explain, the idea that public administration is distinct from politics is an oversimplification of Goodnow's position. Thus, when Goodnow said, "For reasons of both convenience and of propriety . . . the interpretation of the will of the state shall be made by some authority more or less independent of the legislature," he was charging public administrators with the task of giving a truer expression of the public will than partisan legislatures could, an expression akin (though not identical) to Rousseau's idea of the individual will "rightly understood."[58] Consistent with this idea, both Wilson and Goodnow spoke out against popular sovereignty in administrative matters. Wilson, for example, attributed the poor state of public administration during his presidency to popular sovereignty as practiced in the United States,[59] and Goodnow believed that the subjugation of public administration to the principles of popular sovereignty was without theoretical justification.[60] In brief, scholars such as Wilson and Goodnow felt that it was not politics per se that public administration should (or could) avoid. Rather, it should avoid the politics of popular sovereignty and democratic government as practiced in America at that time.

Toward improving governance, both Wilson and Goodnow thought that the processes and practices of public administration were proper subjects for a certain kind of objective study they termed "scientific." As Wilson explained, "nowhere else in the whole field of politics . . . can we make use of the historical, comparative method more safely than in this province of administration."[61] It was supposed that out of such historical, comparative studies there would emerge useful principles to guide the proper structuring of administrative agencies and to direct administrative practices into channels that were accurate, economical, efficient, and fair.[62] As Wilson put it,

> The study of administration . . . is closely connected with the study of the proper distribution of constitutional authority. To be efficient it must discover the simplest arrangements by which responsibility can be unmistakably fixed upon officials . . . If administrative study can discover the best principles upon which to base such distribution, it will have done constitutional study an invaluable service.[63]

Thus, aspiring to reform the American democratic state through the discovery of principles (not laws) derived from historical and comparative studies (not empirical observation in the strict sense), public administration conceived of itself originally as not so much a fundamentally "scientific" endeavor as a fundamentally "social scientific" one.

The distinction is important. Science, insofar as it is empirical, is particularly concerned with the objective and the a posteriori. As a social science, however, public administration is concerned not only with systematically observing, describing, and explaining both recurring social phenomena and habitual patterns of behavior, but with the culture, history, psychology, and purposes preceding and conditioning those empirical "facts" and disposing people either toward those behaviors or toward behaving differently. Particularly, public administration is concerned first with those subjective, a priori experiences, attitudes, values, beliefs, and habits of thought that work toward or against an intersubjective experience of right order and good government. Then it is concerned with fitting institutional practices to those subjective preconditions by inquiring into all that proves significant to the waxing and waning of such an experience in particular contexts. This is one reason why Wilson and Goodnow chose a comparative approach for the study of administration. A comparative approach studies an array of current and historical means to fitting public institutions into collective living that are developed collectively and historically within both a wide range of contexts and an even wider range of purposes, practices, norms, values, attitudes, beliefs, and behaviors. Methodologically, this historical mode of "scientific" knowing employs both micromethods (e.g., case studies) and macromethods (e.g., statistical studies of behavior patterns). It promotes both an anecdotal focus on the idiosyncratic, the unusual, and the anomalous, and a focus on regularly unfolding processes. It emphasizes context, contingency, and uniqueness without committing the category mistake of assuming that the whole is nothing greater than the sum of its parts. In brief, it studies both the whole and the parts, both the transhistorical and the context-sensitivity of general principles to new times and places. Social scientists thus generate alternative ways to think about current issues—both problems and solutions—that might engender an intersubjective experience of good government. They generate, in other words, not laws and unified theories, but workable choices for ways of going on collectively, and they focus attention on recurring problems with certain kinds of choices in certain kinds of contexts that we must reasonably address should we decide to make those choices.

In its earliest incarnation, then, public administration sought to improve governance (i.e., to achieve the intersubjective experience of good government) through "partisan (not political) neutrality."[64] Administrative practices and the scholarship informing them were intended to accomplish enduring and vital public purposes through a thorough commitment to the "public good rightly (i.e., intersubjectively) understood." Toward this end, practitioners and scholars were to rationally focus collective deliberation, to provide and incorporate new knowledge, and to generate rational choices through sound practice, applied knowledge, and self-study. Scholars and "administrators were politicians,"[65] in the sense of "philosopher kings" who did not impose but developed rational choices about how to "go on" in intersubjectively satisfying ways and shifting circumstances, and public administration was not separate from politics but rather a corrective to the politics of policy making and execution in a democratic state.[66] The corrective was administrative processes that were intersubjectively accurate, efficient, and fair as discovered through the social science of a comparative historical study employing both microlevel and macrolevel methodologies. In brief, public administration was defined as a political and social scientific endeavor to correctly assemble and properly operate political institutions to improve the intersubjective experience of good governance.

Even a cursory comparison of this original conceptualization of public administration with Kuhn's characterization of a disciplinary matrix reveals that it constitutes the core matrix of the field that is pointed to by the orthodoxy. That is, it encompasses models of how both the practice and the study of public administration work (simplified in the orthodoxy), formalizable concepts (e.g., public institutions, culturally based goals for collective living), exemplars (e.g., macro and micro social science), and methodological, political, and social values. Moreover, "the shape of the later development of Public Administration—in academe, in the general run of professional journals, and in professional organizations—has done little to break this original pattern"; in fact, "there has been limited pressure against the original form."[67] Consider, for example, the efforts to transform Wilson and Goodnow's distinction between legislative policy making and good public administration, into a general "dichotomy" between politics and administration.[68] Though this attempt at redefinition caught on briefly and casts a shade that troubles us today, it never caught on thoroughly and is heavily discounted as an oversimplification that does not hold up in theory or in practice. Waldo, for example, argued that political theory has guided thought in public administration and concluded that "any simple division of government into politics-and-administration is inadequate."[69] Similarly, Appleby described public administration as the "eighth political process,"[70] and Long pointed out that politics and public administration share the same "lifeblood": power.[71]

Along these same lines, consider the efforts made to conceptually divorce social science from public administration and to introduce more "positivist" values to the discipline (e.g., Gulick and Urwick).[72] This idea that there are universal laws of governance (rather than principles subjectively or intersubjectively crafted) and that it is the job of public administrators to discover and apply them (if necessary in a top-down fashion) proved another unsuccessful attempt to renegotiate the definition of the field by redefining its purposes, values, and methodology. Once again, although the idea was briefly popular it casts a present shadow. As Simon, for example, pointed out, "Administrative decisions mingle fact and value, bring them together in one process or action; and thus there can never be a 'pure science' of administration, in the sense that the value premises or the objectives of administrative action are validated by science."[73] More thoroughly, Dahl directly attacked Gulick and Urwick's positivist renegotiation, noting that public administration must deal necessarily with both "the basic problems of values" and the basic problems of "the individual personality and the social framework," the very (subjective and intersubjective) "stuff" of social science.[74]

Finally, consider the efforts to redefine public administration as a science of voluntary association for collective action.[75] This approach seeks to redefine public administration as a largely bottom-up affair based on the idea that a given population of individuals, each maintaining subjective ideas of good governance, is capable of individually arriving at sufficiently identical subjective goals to voluntarily associate and organize for collective action. Once this occurs, it is thought that individual decision makers will consistently adopt policies sufficiently identical to those desired by the organized group as they will act (1) in their own self-interest, (2) according to the dictates of rationality, (3) with an appropriate quantity of available information, (4) as the requisites of law and order require, and (5) with a view of the least expensive way of achieving preferences.[76] This attempt to redefine the purposes and practices of public administration came under withering criticism regarding its content, terminology, assumptions, methodologies, and implications for practice.[77] To the extent that the idea still holds on, some argue that it is only realistic for small groups. For example, Hardin, Olson, and others point out that given the complexity of conditions necessary for both voluntary association and collective action (i.e., individuals individually acting in their subjective self-interest and yet adopting sufficiently identical goals), there is no reason to believe that voluntary cooperation of this nature is the case in administrative contexts.[78]

For our purposes, the important point about these and other failed endeavors is not that the core definition held against their assault, but why it held. It is our position that the core held because these endeavors attempted to negotiate restrictions upon the scope, methodology, or practices of public administration that were so divorced from the contextual practices and uses of the term as to be unacceptable, if not dysfunctional. In Wittgensteinian terms, the proffered redefinitions bore insufficient family resemblances to the meaning of public administration and therefore did not make sense. Negotiating them into the public administration discourse, then, would have required a paradigmatic shift so wrenching as to render public administration something so completely foreign to what it ever meant before that nothing students or practitioners valued or did (currently or historically) could be included under the term. We might struggle fiercely, for example, to imagine a public administration that is apolitical while pursuing collective goals, securing collective (including constitutional) values and regulating collective behavior. Ultimately, however, it doesn't really make any sense. Equally fierce struggles make equally little sense of the ideas that decision makers need only concern themselves with objective (transcendental) principles of good government, that intersubjective cultural, psychological, and contextual experiences of good governance may be ignored, and that administration need not involve a significant amount of behavior management and control because individuals will, entirely on their own, arrive subjectively at sufficiently identical goals and ideas of how to achieve them.

Evidence bears out the idea that the core has held and that suggestions for change involve extending the core through family resemblances. A 1973 study of the meaning of the term "public administration," for example, revealed "two generally different patterns of opinion . . . among the scholars of contemporary academic public administration: (1) conventional public administration and (2) management and policy sciences."[79] The chief difference between the patterns of opinion, however, was their orientations toward policy impacts and hence toward the scope of researchable topics. Hence, the difference was only in what to research in order to correctly assemble and properly operate political institutions. As the authors indicated, "these findings support the assumption that the alleged 'identity crisis' and confusion in contemporary Public Administration have been overstated."[80]

More recent studies confirm this view. Researching the question "What is 'mainstream' public administration and how has it changed?" Bingham and Bowen content analyzed publications in the *Public Administration Review* and concluded that the "general concerns" of "public administration's theory and substance have changed . . . remarkably little over time." These general concerns were focused on "government and organization behavior, public management, and human resources," all clearly cleaving to the "original [definitional] pattern."[81]

More definitively, Lan and Anders content analyzed "mainstream public administration journals" in order to "shed light upon" their concerns that there may be "no viable, broad-ranging paradigm to govern public administration research" and that there may be no "intellectual core for the field of public administration."[82] Pointing out that according to Kuhn, "competing paradigms can exist within one discipline,"[83] Lan and Anders concluded that,

> . . . the field of public administration could claim to have its paradigm . . . this paradigm asserts that public administration differs from other types of management, private management in particular, in meaningful ways . . . [and that] the objective of the normal science research within this paradigm is to improve public service performance and resolve various problems encountered in public administration practice.[84]

Further,

> Under this paramount paradigm, competing approaches exist. These approaches, too, qualify for Kuhn's definition of paradigms. Because these

paradigms are conditioned by and identified within the boundary of the major orientation of the field, they may be referred to as the second-tier paradigms or cognitive approach subparadigms. The paramount paradigm defines the boundary and legitimate issues for the field, whereas the cognitive approach subparadigms define the conceptual, theoretical, and methodological approaches used in the research within the field.[85]

These subparadigms include

> (a) the political approach, ... [or] public administration ... as policy making, power struggle, and resource allocation ... ; (b) the managerial approach, ... [or] public administration as an instrument to achieve social and organizational efficiency; (c) the judicial approach, ... [or] public administration as an instrument ... upholding the Constitution and other ... laws and regulations; (d) the ethical approach, ... [or] public administration ... [as]cognizant of ... its impact on democratic values such as liberty, justice, and human dignity; (e) the integrated/comprehensive approach, [or] public administration as ... [doing] whatever necessary to keep the government functioning; and (f) the historical approach, [or] public administration as defined by trackable historical contributions to the theoretical development of the field or to the practice of public administration.[86]

For our purposes, two things are most important about these findings. First, Lan and Anders uncovered neither a heuristic public administration nor a set of paradigms clustered together under the rubric public administration. Rather, they uncovered the persistence of an entire "disciplinary matrix." This is most important as the existence of a disciplinary matrix, in Kuhnian terms, delineates public administration as a distinct "scientific community" (i.e., as a scientific discipline). Second, the revealed disciplinary matrix does not break the original definitional form.

As to the first point, recall that seeking to elucidate the various uses of "paradigm" within his original theory, Kuhn explained that there are paradigms in the narrow sense and paradigms in the fullest sense.[87] Paradigms in the narrow sense are best thought of as only one element in the definition of a scientific discipline. Disciplines are particular communities united by a disciplinary matrix (a paradigm in the fullest sense). This disciplinary matrix includes symbolic generalizations (formal or readily formalizable components), models (deliberate simplifications of reality), values and exemplars (paradigms in the narrow sense or concrete problems that can be solved by the methodological forms provided by the community).[88] The dominant paradigm/subparadigm complex revealed by Lan and Anders as operative in public administration appears to satisfy each of Kuhn's requirements for a disciplinary matrix. For example, Lan and Anders point out that the public/private generalization has been formalized in a significant body of research, as have certain useful models.[89] The "ethical approach" subparadigm and the commitment to sound social science reveal the value dimensions of the dominant paradigm, and exemplars of such research are identified for both the dominant paradigm and the subparadigms.

As to the second point, subparadigms elucidating the "political endeavor" of public administration (subparadigms a, c, d, and e) do not break the form but elaborate and deepen it. Struggling for power and resources, putting law into action, regulating behavior according to collective (intersubjective) norms, safeguarding democratic values, and sustaining government are all political endeavors. Researching these forms of political endeavor can "improve public service performance and resolve various problems encountered in public administration practice."[90] It certainly seeks to correctly assemble and properly operate political institutions so as to improve the intersubjective experience of good governance. Subparadigms elucidating the operation of political institutions (subparadigms b, d, and e) similarly advance what is contemplated by the "definitional form," and studies of historical contributions reinforce the idea that what counts as public administration at any given time is the result of a discourse grounded in the original definition and extended or deepened through negotiated change.

Conclusion

Overall, then, we might comfortably say that public administration is in fact a distinct discipline (i.e., a coherent disciplinary matrix) defined by its publicly political and social scientific endeavor to correctly assemble and properly operate public institutions so as to improve the intersubjective experience of good governance in practice. Furthermore, we can say that the discipline as so defined, though constantly under negotiation, has never really lost its identity. It retains its core definitional complex of purposes, practices, values, and behaviors, though it now intelligibly includes peripheral complexes bearing family resemblances to the terms of the core. Furthermore, we may conclude that because public administration has a distinct identity, it is capable of developing a theory of its own. The successful renegotiations at the periphery of what counts as public administration reflect only the fact that what inclines people toward an intersubjective experience of good governance in practice is changing as governing becomes intertwined with both the society and the individual to a degree not originally contemplated.

Unfortunately, the mere fact that public administration has an identity as a social science of generating in practice the intersubjective experience of good governance is insufficient to garner it a theory. To develop a meaningful, nonillusionary theory, the subject matter of public administration must have ontological status. Theories about unreal things are myth and mirage, and they do not elucidate the nature of the discipline. Therefore, we must next understand the ontology of public administration.

3

Ontology and Theory in Public Administration

Although public administration constitutes a distinct discipline defined by a distinct disciplinary matrix, many still argue that any theory of public administration must remain myth and illusion. This is so, the argument goes, because social phenomena, behavior, and discourse are conceptual and conventional in nature rather than the sort of brute data that is the subject matter of the natural sciences.[1] Given the nature of this subject matter, the job of theory is not so much to describe and explain as to interpret the conventions and assumptions about reality and knowledge lying behind social phenomena (i.e., theory must be hermeneutical and not scientific). More particularly, because the subject matter of public administration is not composed of brute data, it lacks any tangible reality beyond the conventional practices, behaviors, and discourses trackable to its initial use. As these conventions alone are the proper subjects of normative and empirical theory, public administration (a mere artifact assembled from them) has no ontological status and cannot itself be the subject of normative or empirical theory.[2] In brief, public administration as a field lacks ontological status because it is among those "activities that study other activities . . . [and] that cognitively confront a subject matter that is discursively preconstituted and preinterpreted . . . [rather than] theoretically constructed by social science."[3] Public administration theory, then, remains theory; it is about the constructs and not about anything real.

We contend that this thesis is in error and that the subject matter of public administration has ontological status. Moreover, we will argue that the ontological nature of public administration's particular subject matter requires that its normative and empirical theory be critical and evolutionary in nature and develop through a synergy among institutional, behavioral, and hermeneutic methodologies. In this chapter, we will present our argument that public administration's subject matter has ontological status and that, given the nature of that subject matter, theory in public administration must develop through a synergy among methodologies. Next, we will elucidate and refine our argument by working to untangle certain knotty problems that it entails. Finally, we will introduce certain core social and political concerns that must be addressed by public administration if its claim to cognitive authority regarding its subject matter is to be honored. In the following chapters, we will endeavor to demonstrate the advantages that an evolutionary critical theory provides as a necessary means of effecting synergy among institutional, behavioral, and hermeneutic methodologies and thus addressing public administration's core concerns.

The Ontological Status of Public Administration Theory

The core problem with the argument denying ontological status to both public administration and its theory is that it misconstrues the nature of public administration's subject matter. While many social sciences (e.g., political science, sociology) are concerned with various and particular social phenomena and social behaviors, public administration is concerned with these only insofar as they as embody, reflect, advance, or retard the intersubjective experience of good governance. More particularly, public administration is concerned first with those subjective experiences, attitudes, values, beliefs, and habits of thought that work toward or against an intersubjective experience of good governance, and then with fitting institutional practices to those subjective conditions by inquiring into all that proves significant to the waxing and waning of such an experience in particular contexts. To the extent that these subjective and intersubjective experiences are its subject matter, public administration and its theory are true to what has for a very long time been understood as reality and theory.

In the ancient Greek, for example, *theoria* refers to the state of heightened awareness, intense involvement, and emotional purity experienced by spectators during the public performance of a tragedy.[4] This reference derives from the fact that *theoria* was employed initially in the context of visiting religious festivals as a spectator.[5] Later, the term was employed in any context wherein a person attempted to pierce the veil of appearance in order to perceive reality directly rather than through *doxa*.[6] The common referent of its use in any context, however, was the attempt to perceive "something outside the observing self" that was subjectively "relevant to one's own emotions, needs or desires."[7] Good theories thus "discriminate and ... discern" from "an otherwise chaotic background" by defining and forefronting objects and patterns of activity that are useful in addressing subjective needs, interests, and desires.[8] Moreover, the context, the objects, or patterns forefronted, and the importance or relevance of what is discerned, "must be capable of being verified, directly or indirectly

... that is, either corroborated or disconfirmed."[9] That is, good (truly rational or practical) theory requires intersubjectivity with respect to both the discernment of the objects or patterns forefronted and the experience of their relevance to individual needs, emotions, and desires.

Now, the intersubjective experience of relevance that is vital to *theoria* is no less real than either the objects and patterns it plays a part in revealing, or the "stuff" or brute data of the *real world* from which such objects and patterns are discriminated. True, experiences are to some significant extent more culturally and historically determined than the data from which they are abstracted (though much of that "stuff" might be socially constructed as well). And they are arguably somewhat more dependent for their existence upon tradition, the relationships of ordinary life, established discourses, and institutional structures. However, this only means that such experiences (realities) are ontologically "less settled" and cognitively less certain than a strictly positivist ontology might anticipate. Methodologically, it means only that "discovering" and "articulating" these "background" experiences and the "frameworks" within which they occur are "interwoven with inventing them."[10] And theoretically, the ontological unsettledness of the particular reality of concern, and the diminished claims to cognitive certainty that may be made as a consequence, only mean that the persuasiveness of one articulated set of experiences (one theory) as opposed to another will rest finally upon both its adequacy with regard to the intersubjective experience of good governance and its appropriateness to the intersubjective experiences (themselves contestable) of the historical, cultural, and institutional context as well. This, of course, indicates the importance of actively, continuously, and experimentally bringing those experiences into continual, practical engagement with one's surroundings and ensuring that the subjectively experienced results are articulated in some intersubjectively meaningful way.

Implications for Theory

Briefly, then, we do not experience the world but only have experiences of it, and these experiences are real despite their transitory and sometimes conflicting nature. Consequently, the conventional practices, behaviors, and discourses that some say can be the only proper subjects of theory are in fact neither ontologically nor experientially given but generated and known only through conceptual systems, theories, and (often tacit) metatheories. Moreover, they are distilled to manageable proportions primarily through interpersonal discourse in a range of languages (e.g, scientific, religious, historical, legal) that are constructed according to particular interests and purposes arising from experience.[11] Thus, social scientists focusing on these artifacts for theoretical purposes "confront a subject matter that is epistemologically and discursively pre-constituted and pre-interpreted"[12] by the individuals, groups, and institutions they study, and consequently not the proper subjects of theory. Understanding these phenomena necessarily "involves a dialectical relationship between two social constructions [two theories of experience]—that of the social scientist and that of the social actor."[13] Given this, constructing an ontologically based (truly rational and practical) theory of social phenomena involves pushing beyond this dialectic to accomplish an identity with, sharing of, or at least an intuition of the experiences giving rise to the ideas, intentions, preconstructions, values, and discourses behind individual and group behavior and discourse. That is, social scientists must integrate theory, practice, and intersubjective experience to provide a theory with ontological status that is speculative, fluid, and practical simultaneously.

The nature and ontological status of public administration's subject matter, then, requires that its theory display several distinguishing characteristics. First, its theory's central concern must be the transition points or loci of synergy between the individualistic and the global. As is the case with all social science, public administration is concerned not only with recurring social phenomena and habitual patterns of behavior, but also with the culture, history, institutions, psychology, understood meanings, and shared purposes of such behavior and phenomena. But what distinguishes public administration theory is the fact that because its central concern is how the intersubjective experience of good governance affects and is affected by administrative institutions, it must focus upon the crossing points between the individual and the social, between purpose and phenomena, between meaning and behavior, between institutions, individuals and groups, between culture and psychology, and between the particular and the general.

This requirement is profound in the sense that all other requirements follow from it. For example, because it focuses upon these crossing points in order to understand how the intersubjective experience of good governance affects and is affected by administrative institutions, public administration theory must be wholly embedded in an institutional perspective. It must look at the individual and the global, the recurring and the contingent, the particular and the universal, but always with a view to fitting institutional practice to all that proves significant to the waxing and waning of an intersubjective experience of good governance. In addition, public administration theory must seek to explain both individual and aggregate behavior and choice (both "irrational" and utility maximizing) as the result of not only individual experience, but also (1) the autonomy (agency) of administrators as they actively seek to affect aggregate behavior, secure institutional interests, and pursue institutional goals by, inter alia, manipulating important symbols and structuring institutions and their processes, (2) the institutional rules, roles, norms, and the expectations that constrain individual and group choice and behavior, and (3) the conditioning of individual, group, and institutional identities, values, and habits of thought by history, culture, context, and roles vis-à-vis institutions and their needs. Moreover, public administration theory must encompass both historical forces working for social and institutional equilibrium (negative or positive) and those forces (contextual and historical) rendering the intersubjective experience of good governance elusive and hence working toward breakdown.

To accomplish this, of course, theorists of public administration must "relax all preconceptions as to how the subject matter might behave, and permit themselves to be instructed by the subject matter"; to "allow the subject matter to impress upon them its norms and to instill within them its categories."[14] Only after they have done this should theorists "draw back to become objective and struggle to formulate a patterned explanation of the behavior that they have come to experience,"[15] and then only momentarily. Finally, public administration's subject matter requires that its theory embrace a methodology that can serve as a nexus among institutional, behavioral, and hermeneutic approaches, because all of these are required both to fulfill the above requirements and to clarify the intersubjective experience of good governance, how it comes about and how it is lost.

Implications for Methodology

Given these distinguishing characteristics, public administration must proceed methodologically by developing a synergy among institutional, behavioral, and hermeneutic approaches and methodologies. It is insufficient for public administration to recognize the strengths and weaknesses of each and to employ each where most appropriate, because its subject matter occurs at precisely the point where institutional structure, process, and power meet meaning, values, and behavior. Hence, employing these methodologies independently or only in traditional ways renders them less useful than does recognizing that while they were logically and culturally constructed for certain purposes, they may be used in different ways and reconstructed to some extent to serve other purposes. Thus, public administration must seek a synergy among the three in order to secure the range and depth required of a practical, speculative, and fluid theory explaining how the intersubjective experience of good government waxes and wanes. This means, of course, that (1) theory testing toward the development of global covering theories, (2) the stepwise development of middle-range theories "adequate to limited ranges of social data,"[16] (3) more localized inquiries (e.g., situational analyses, qualitative research, and action research) that add depth and nuance through a detailed consideration of the natural and social environments and the lifeworlds of individuals, and (4) hermeneutical approaches to understanding the historical and cultural meanings and prejudices of institutions, groups, and individuals are equally important modes of inquiry, and public administration must devise a way of making them work in concert (in an interactive and mutually developing way) toward an understanding of its subject matter.

In sum, the methodology of public administration must address the following four ideas. First, it must take as objects of study the intersubjective experience of decision makers, groups, and citizens in general and how institutional structures, rules, norms, cultures, and processes fit to those experiences. Hence, it must employ both survey research and the institutional approaches in such a way that public administration theory might integrate the results. In the process, inquiry should not focus exclusively upon administrative institutions or behavior in general, but upon specific institutions and specific groups exhibiting (or not exhibiting) specific types of behavior. Second, it must deepen the level of explanation of intersubjective experiences of good government by employing hermeneutic approaches and localized methodologies to understand the meaning and nuances of individual, group, and institutional phenomena, goals, values, behavior, and choices, to those acting and choosing, those affected, and those observing actions and effects. Thus, it provides for a variety of social, political, and economic explanations and evaluations of and beyond, (e.g., utility maximization, heuristics, and bounded rationality) that provide for the possibility that administrators and citizens alike are neither unreasonably rational nor cognitively discomfited. Third, it must pay close attention to the environment (through various forms of qualitative research, situational analyses, and action research), not only the political environment but the lifeworlds of those affected substantially by administrative institutions and the social, cultural, legal, and economic environment as well in order to understand both the broad and the local factors impinging upon the intersubjective experience of good government. Finally, it must devise a procedure for integrating synergistically the various methodologies employed in order to shape the results of its inquiry into dynamic models capturing the synergy among behavior, meaning, values, and institutional factors as they affect the flux and evolution of intersubjective experience and institutional change.

Problems

Certain knotty problems are entailed both by our suggestion that the nature of public administration's subject is the intersubjective experience of good governance and by our delineation of the distinguishing characteristics of its theory and methodology. These problems go to both the core of its subject matter and its implications for study and practice, as they question whether what we have proposed so far is analytically and empirically sound, theoretically understandable, internally consistent, and consistent with what we know about the nature and workings of the natural and social worlds. Consequently, we must address these problems before arguing the advantages of a critical evolutionary theory given the nature of public administration's subject matter and the requisites of its theory and methodology.

The Problem of Intersubjectivity

Perhaps the most fundamental problem entailed by our contention that public administration's subject matter has ontological status involves the claim that intersubjective experience is no less real than either the objects and patterns it plays a part in revealing, or the "stuff" or "brute data" of the real world from which such objects and patterns are discriminated. According to our argument, the ontological status, rationality, and practicality of public administration theory depend ultimately upon whether there are such things as intersubjective experiences and whether those experiences are of a common subject matter. Absent a systematic intersubjectivity (i.e., without the experience of a common subject matter that is known by at least two people to be common), there is no potential for a shared reality, and without a shared reality there can be no common development or recognition of the concepts, categories, and principles necessary to both cognition and empirical theory. In addition, our argument is correct only to the extent that we can delineate how what we experience intersubjectively comes to have shared meanings. Absent shared meanings, there is no ground for evaluating our shared experiences and we cannot theorize normatively. Consequently, to establish that there can be normative and empirical theory in public administration, we must meet the assertion that each of us lives in a private world, a world unknowable to others, a world without shared experience, and a world we cannot communicate meaningfully.

Do individuals share the experience of a common subject matter (say, an external natural world) that they know to be common and about which they can communicate? And if they do, how is this possible? To begin answering this question, we should recognize that if we are living in private worlds there can be no intersubjectivity (no sharing of any thought or experience). Assuming for the moment that it is possible for each of us to have subjective thoughts and experiences in such a private world, there is no assurance that I am experiencing or thinking anything that you are experiencing or thinking. Hence we can communicate only by happy accident, because whatever gestures I might make (verbal, written, or physical) have no shared meaning and no demonstrable grounding in your experience. However, merely stating this idea involves a paradox; otherwise, to whom are these sentences addressed, and for what purpose? To mutually consider the question of intersubjectivity, in other words, requires the enabling condition of intersubjectivity. Moreover, for the intersubjective experience of discussing intersubjectivity to make sense, we require some common language game, and that must come from somewhere external to either of our private worlds (i.e., either we must intersubjectively establish it or receive it intersubjectively from a third source).

So intersubjectivity seems inescapable, but is that intersubjectivity the result of experience with a common subject matter we know to be common or from something else? As we cannot experience the world but only have experiences of it, our cognition cannot be of things in themselves. Hence, an external world cannot provide objective points around which our individual experiences might coalesce. As the world in itself cannot be confronted, any intersubjective experience of it must be either ontologically given (transcendentally or mundanely) or contingently accomplished by our historical and contextual practices.[17]

The idea that intersubjectivity is given transcendentally is ultimately a matter of faith. According to this view, the external world is intersubjectively knowable because both the world (a common, external subject matter) and everybody in it participates in an absolute reason or underlying unity (perhaps but not necessarily flowing from a transcendental being) that is an actuality in every act of cognition. It is this reason or unity that constitutes a synthetic apriority allowing us to make intersubjective sense of a common subject matter. Now, this transcendental apriority is clearly posited as enjoying ontological (though perhaps metaphysical) status, and as such it should be accessible to experience, and that experience should be the same for everyone, given its nature as a synthesizing agent. Unfortunately, claimed experiences of it are notoriously varied, as is the sense made of both it and the world it is claimed to clarify.[18] Thus, as there seems to be no intersubjective cognitive access to an underlying apriority, its ontological status is problematic and it remains at best a practical or useful postulate whose existence must be assumed or imagined.

To avoid such acts of faith, many point out that the argument favoring a transcendentally given intersubjectivity presupposes that intersubjectivity can be understood only if we start with entirely discrete individual subjectivities and proceed to explain how such discrete subjectivities bridge the gaps among them. However, these people argue that there is in fact no convincing evidence for the existence of such wholly discrete subjectivities and that we all experience the development of our individual selves from (and perhaps eventually in contradistinction to) others. For this reason, we (and everything else "in the world") become "discrete" only "in correlation with and through [others who] . . . reveal me [and the world] to myself."[19] More broadly, people perceive both themselves and a shared external world only upon being furnished by those around them with commensurate conceptual systems, epistemologies, and worldviews that "construct" the experience of a "self" and a common external subject matter.[20] Once this is accomplished, we might confidently conclude that "the world is not my private world but an intersubjective one . . . therefore, my knowledge of it is not my private affair but from the outset intersubjective and socialized."[21]

Unfortunately, while this socialization approach shifts the source of intersubjectivity from the transcendental to the more material level of social interaction, it does not by itself answer the question of how the actual socialization into the " reciprocity of perspectives" is accomplished.[22] The argument in its most stringent form seems to be that intersubjectivity emerges from the subjective experience of a set of relatively autonomous conventions (e.g., socially constructed conceptual systems, epistemologies, and worldviews). Thus, because the capacity for intersubjectivity is not "located" originally in the "self" (i.e., because the "self" and its capacities are socially constructed), and there can be no prelinguistic or presocial experiences of any subject matter at all, the experience of intersubjectivity can result only from particular congeries of relational influences operating on a largely "empty" subject.

But what does this mean? Does it mean, for example, that the experience of intersubjectivity is "relative to one theory or body of assumptions" (i.e., the group's set of data, experience, and conventions) and that it may fail to emerge absent of such assumptions?[23] While it seems to make sense to talk about certain kinds of phenomena (e.g., a personal identity, a conceptual scheme, a value system) failing to occur under the conditioning impact of particular social conventions, the idea seems to founder with respect to intersubjectivity. Would we really fail to experience intersubjectivity if we were socialized differently? Moreover, absent intersubjectivity, how could we be socialized at all? How is it possible, for example, to conceive of a person embracing a culturally constituted identity or conceptual system without referring to intersubjectivity? Similarly, into what sort of assumptions could we be socialized so as to avoid intersubjectivity? And where would such assumptions come from? Must not they have an intersubjective source? And if they have such a source, how does *that* intersubjectivity emerge? At this point, either we are caught in an infinite regress or we must conclude that intersubjectivity is a preexisting enabling condition for transmitting conventions, discourses, and conceptual schemes. In brief, socialization theory cannot adequately explain the phenomenon of intersubjectivity because socialization is possible only if there is already a less contingent intersubjectivity (experience of a common subject matter) already in place. And as it is less contingent, intersubjectivity must be grounded in something other than the immediate social or physical context, existing independently of our specific symbols, interests, discourses, institutions, habits of thought, and social interactions.

Thus intersubjectivity seems unexplainable in terms of either transcendental agents or socialization processes that produce both a self and a common subject matter. Consequently, it cannot be externally grounded and we must commit ourselves to the view that intersubjectivity has no external cause at all but only an external subject matter. That is, as intersubjectivity does not emerge as a generalized intersection of perspectives generated by either a transcendental unity or the ramification of social interaction, the capacity for intersubjectivity must be already ontologically constituted and located not transcendentally or socially but in the self; and as the activator of that capacity can neither be identified transcendentally nor socially constructed, there must be a common subject matter being experienced. And as only an ontologically constituted being can be rendered cognitively self-conscious (i.e., a self must already be in place before any attempt at transcendental unity or socialization may proceed),[24] the self must be the mundane ground or originating location of both intersubjectivity and an ontologically constituted capacity for intersubjectivity. In brief, we must, as human beings, be so ontologically constituted as to be capable of intersubjectivity, and intersubjectivity must be ontologically constituted as a human function or aptitude.

Despite the mundane ontological givenness of intersubjectivity, many still try to explain it in causal terms. However, these attempts are not only self-contradictory but futile as well. They are self-contradictory because an explanation of an ontological given suggests that it is caused and not given; and they are futile because they require the postulation of other givens that must be assumed or imagined. For example, many hold that as the "real world" is in fact unknowable, a synthetic apriority is ontologically given as the mind's contribution to the ordering of our experiences. This apriority is thought to be less abstract than a universal reason or underlying unity when it is posited as being located in the mind, the experience of which we all enjoy. Along this line, Kant argues that the form but not the content of intersubjective experiences is ontologically given since direct experiences of the world are mediated by delineated categories in the mind. Thus, a workable intersubjectivity is accomplished when whatever differing content two people might experience is contained or "fitted into" shared mental categories. However, because these apriority categories themselves cannot be experienced, a leap of faith as to their existence is required.[25]

Similarly, Husserl proposes that we bracket-out presuppositions (theories and assumptions) about both external objects and the modes of experiencing them.[26] He proposes instead a "phenomenological reduction or suspension of such presuppositions in order to take us" back from the hitherto naively accepted world of objects, values and other men, to the transcendental subjectivity that "constitutes" them.[27] That "transcendental subjectivity" or "transcendental Ego" is the "self" or consciousness left over after common assumptions are "reduced" or suspended.[28] But to the extent that this "self" is "transcendental" (in the sense of being shared with others), it must be a theoretical construct. If it were more than theoretical, surely we could each have the same experience of it, and it would thus be that part of the "naively accepted world" that we would all point to in defining objects and selecting values. Moreover, as experience is necessarily a posteriori, how is it possible for an a priori "transcendental Ego" to experience itself? That is, how could it be ontologically constituted if it must constitute itself by means of itself? In brief, such an ego must also be at best a practical or useful postulate whose existence must be assumed or imagined.

Thus, we are returned to the conclusion that intersubjectivity seems an irreducible and nondeducible "datum of *everyone's* lifeworld."[29] What is more, as "the possibility of reflection on the self, discovery of the ego . . . and the possibility of . . . establishing a communicative surrounding . . . are founded on the primal experience of [intersubjectivity]," it must be both "the fundamental ontological category of human existence . . . and . . . the foundation of all other categories of human existence."[30] In brief, intersubjectivity is not only the experience of something real (otherwise to whom are these sentences addressed, and for what purpose?), but *the* primal and irreducible experience.[31]

The Problem of Divergent Intersubjective Meanings and Particular Particulars

Now, if the experience of (and aptitude for experiencing) a common subject matter that is known to be common is the primal ontological given, how are different understandings of it possible and how can we err in perceiving its contents? Do not our errors and varied perceptions indicate that we do not really share a common subject matter? Stated from a more theory-oriented standpoint, "Why is theory necessary?" Given the primacy of intersubjective experience, how is it that understanding the thoughts, perceptions, and behaviors of others becomes so problematic as to require empirical and normative theory to be understood and evaluated? Does not the very process of theorizing indicate that our intersubjective experience of a common subject matter is either mere hypothesis or perhaps a form of provincialism?

As we have argued, the experience of such a shared subject matter is so common as to be unremarkable and so to go unnoticed. It is, for example, the unremarkable presupposition of all empirical science and scientific theory. The experience of an external "nature" as a common subject matter is everyone's experience; and the objection that this is merely a hypothesis or a form of naive provincialism not shared by others founders on the fact that those making such a suggestion are prompting the intersubjective experience of another view and relying upon the intersubjective experience of particular views as being wrong about something external (i.e., they again presuppose a preexisting intersubjectivity of some subject matter). The point is that even the suggestions of errors and alternatives point to an ontological intersubjectivity necessary for the spread of any idea. Moreover, these suggestions throw the dimensions of intersubjectivity into sharp relief by highlighting the general and particular propensities of intersubjective experience. That is, because intersubjectivity is the experience of a common subject matter that is known by at least two people to be common, it involves necessarily the common experience of at least two "particulars" (the self and another) and the common experience of a general subject matter (e.g., the external world or nature). Consequently, all that differing perceptions and error indicate is the nature of primal intersubjectivity; it is the experience of (and aptitude for experiencing) both the common and the particular.

The real question posed by differing perceptions and error, then, is not whether intersubjectivity is a primal aptitude of the self, but how we arrive at differing intersubjective meanings and how we come to distinguish different intersubjectively particular particulars (objects and patterns) from a common subject matter. It is obvious, for example, that we have intersubjective experiences of (and hence an aptitude for experiencing) intelligible unities. That is, our "cognitive acts . . . are not isolated particulars, coming or going in the stream of consciousness without any interconnections. [Rather], they display . . . corresponding connections . . . that present an intelligible unity."[32] Thus, we regularly distinguish or forefront certain objects and patterns (particular particulars) from our constantly changing, sometimes interrupted and unstable experience of the commonly experienced external world. Likewise, though these experiences of relatively stable and intelligible unities are subjective and relative, we "normally in our experience and in the social group united with us in the community of life . . . arrive at 'secure' facts."[33] That is, despite the experience of (and aptitude for experiencing) both a common subject matter and the particular, different groups share different (and fluctuating) subjective experiences (differing unities).[34] Moreover, different groups share different meanings of their experience (e.g., there is something to fear or love, the outcome is just, the person is good). How is it, then, that different groups share the experience of a common subject matter but different experiences of particular particulars? And how is it that they give those experiences different meanings?

The answer to this question lies in the socialization process, operating both generally and locally. There is convincing evidence, for example, that our aptitude for experiencing both a common subject matter and the particular is channeled into culturally acknowledged ways of seeing and understanding by common socialization processes.[35] General or broad socialization processes are the mechanisms for arriving at broad intersubjective meanings and for distinguishing particular particulars in a broadly shared way. Beginning in childhood and excluding or marginalizing groups or subcultures whose worldviews, values, attitudes, and beliefs diverge substantially from those of the dominant culture, broad socialization processes furnish us with not only commensurate conceptual systems, epistemologies, and worldviews but commensurate normative systems as well. People thus begin to both perceive and evaluate the world in roughly the same way. Consequently, both the empirical and the normative content of our minds tend to converge,[36] not according to any demonstrable matching with an ontologically constituted apriority, but according to the experiences of others within our historical, institutional, and cultural contexts.[37]

But socialization operates on another plane as well, and it is experience on this plane that accounts for differing perceptions, assorted conceptual schemes, a range of epistemologies, and error. At this level, we learn practices and techniques for operating successfully within particular localized contexts, discourses, and institutions that compose our concrete day-to-day lifeworld. These learned, practical, adaptive behaviors may vie with historically and culturally generated ways of seeing and doing, thereby accomplishing both a certain divergence of intersubjective meanings and the forefronting of different particular particulars.[38] Localized intersubjectivities thus arise through our contextual maneuvers within the context provided by historical, cultural, and institutional attitudes, values, and beliefs into which we are socialized, and make possible divergent meanings, different ways of seeing particular particulars, and the construction of novel particular particulars as well.

Divergent intersubjective meaning and differing particular particulars, in other words, originate in and develop from the conjunction of our broad socialization and our mundane, moment-to-moment practical socialization in the immediate context. Particular particulars and particular meanings are developed in response to something outside the observing self that is intersubjectively relevant to our interests, emotions, needs or desires." These relevant particulars are developed into particular "relevance structures" apropos our historical experiences; and they are erected into an overall "form of life" that is the background, making it possible to communicate meaning in a localized interpersonally experienced world. Once constructed, these shared presuppositions, common habits of thought, and conventional patterns of behavior are "socialized into" us and modified (temporarily or permanently) through daily practice. In brief, we arrive at differing intersubjective meanings and come to distinguish differing particular particulars through socialization to both a collection of historical and cultural practices, values, and institutions, and a collection of practical behaviors provided by a variety of context-specific practices encountered and tested in our immediate "lifeworld."[39] At neither level, however, are the resultant shared meanings or the intersubjective perceptions of particular particulars the result of any convergence between multiple subjectivities; rather, they are phenomena cooperatively and conflictually produced among socialized individuals. And it is these conflicts and cooperations that account for differing perceptions, assorted conceptual schemes, a range of epistemologies, and error regarding the common subject matter we all experience.

The Problem of Intersubjectivity and "Identity Thinking"

Perhaps the most important problem for public administration as both a discipline and an endeavor is broached at this point in our argument, and it is a problem not with the argument itself but with its implications. As argued above, broad socialization into shared concepts and epistemologies is necessary to both meaningful thought and meaningful communication at both the broad and the local level. However, one potential effect of this socialization is particularly troublesome: the tendency to identify (equate) objects, subjects, and situations with the concepts developed as helpful tools for thinking and communicating about them.[40] So long as it is clear that concepts are tools and that we may need to retool whenever particular concepts fail to serve us well, the current interests, needs, and desires that are intersubjectively recognized at the local level stand a good chance of being formulated in a meaningful (rational) way, communicated effectively to the broad level of society, and addressed by society in an equally meaningful and rational way. The problem is that once identity is achieved, people at both the broad and local levels tend to attribute the properties of the concept (real and imagined) to the subject, the object, or the local situation. The result is often a denial of or blindness to both local realities and the actual nature of the subject, its interests, needs, and desires. Once this occurs, public administration's goal of attaining an intersubjective experience of good government becomes problematic.

As an example, consider critical theory's analysis of how religion functions. Early on, critical theorists recognized that insofar as religion masks its authority structures, obligatory practices, and particular epistemologies as reifications of universal truth, it imposes a collective subjectivity from which people must be emancipated in order to recognize their true interests.[41] Operating in this manner, religion loses its ontological foundation in intersubjective experience and seeks to marginalize or eradicate the articulation of those intersubjective experiences (theories of reality) that are not in accord with its ideology. As emancipation requires demystification and the empowerment of individuals and groups through the transformative fusion of theory and practice,[42] religious impediments to such an emancipation should be either removed (the original Frankfurt theorists) or "renegotiated" in accordance with either proper communicative practice (Habermas) or the normative presuppositions of social interaction (Honneth) whenever they subject individuals or groups to marginalization, disrespect, or denigration.

However, critical theorists also recognize that religion may operate in an ontologically grounded way. Adorno, for example, understood that "religious authority" may constitute an articulation of the intersubjective experience that is "not mere domination; like the traditional authority embedded in historical values and hierarchically structured power relations, but the product of communal negotiation of needs in response to the threats of nature."[43] This particular form of authority, argued Adorno, plays an emancipatory role by guaranteeing "key elements of psychopolitical autonomy."[44] Thus, "while, individual religious contents serve the purposes of traditional authority and domination . . . religion also contains a universal feature of all societies as the communicative precondition for knowledge of the world. Religion . . . in Adorno's thought, thus occupies a paradoxical space as both form of domination and locus of truth."[45]

Similarly, Habermas understands that intersubjective forms of religion may constitute a defensive reaction to the penetration and "colonization" of individual "lifeworlds" by the increasingly autonomous practices and epistemologies of economic and political institutions.[46] Thus, while, "forms of religious dogmatism" may thwart "discursive will formation" and create certain psychopathologies (e.g., loss of meaning, crises of anomie, and crises of motivation), "religion as an intersubjective response to contradictions within social systems and to struggles among the state, economy, family, and [dogmatic] religion" may encourage emancipatory efforts.[47] Religion and its institutions, in other words, may operate as either a classic form of domination or an expression of current intersubjective experience. As the product of a communal negotiation of needs in response to external threats (e.g., from nature or from the imposition or insinuation of foreign values, attitudes, or beliefs), religious dogma can manifest shared interests rather than interests imposed by a dominant ideology.[48] Hence, empirical findings suggest that "only when religion does something other than mediate between man and God does it retain a high place in people's attentions and in their politics."[49]

Just as religious concepts, epistemologies, and institutions can operate either as forms of domination or as communal expressions of intersubjectively recognized interests and desires, all conceptual schemes, epistemologies, and institutions operate potentially in both modes. Administrative institutions, then, can hope to attain and retain an intersubjective experience of good governance only so long as the individual contents of their conceptual schemes and epistemologies are recognized as particular particulars endowed with particular meanings that are built up into particular relevance structures that become dysfunctional when they no longer serve. And we can tell when they no longer serve by attending to whether they currently thwart intersubjective will formation, as at that point they become dominating institutions acting according to bad theories that define and forefront objects and patterns of activity that are no longer meaningful or useful in addressing intersubjective needs, interests, and desires.

To sum up, truly rational (meaningful) relevance structures (conceptual schemes, epistemologies, and values) are generated interpersonally as useful tools apropos our recent experience with the brute data of a commonly experienced subject matter, the necessities of our lifeworlds and the demands of the cooperative roles that we occupy in the total social structure. What constitutes a useful tool, however, changes according to our experience with each of these three factors. Consequently, particular relevance structures are at best relevant to our historical experiences and only hypothetically relevant to our current set of intersubjective experiences. Administrative institutions, then, as embodiments of these relevance structures, can expect to engender the experience of good governance only so long as they change to meet these experiential changes. Experiences of bad governance, on the other hand, arise when administrative institutions adopt one historical relevance structure (or some utopian ideal) as true across time and space (e.g., when they adopt one conceptual scheme as more than a discardable tool and equate their historically useful conceptualizations with reality), thereby divorcing themselves from current experiential intersubjectivity and engendering the alienation and marginalization of those whose experiences cannot be fitted into the adopted relevance structure. Thus, as Foucault suggests, identitarian gestures by which the other is subjected to discipline are what public administration must guard against most of all.[50] Public administration, then, cannot rely upon identity thinking to produce its desired result; it requires a theoretical and empirical approach to its subject matter that obliges it to constantly seek out and understand localized intersubjectivities. It is only through the critique of both those intersubjective understandings and its own currently fundamental concepts, values, assumptions, and habits of thought that it can best determine how local and more broadly shared experiences might be negotiated into the broadest possible intersubjective experience of good governance.

The Problem of Irreconcilable Intersubjectivities

Of course, attaining the broadest possible intersubjective experience of good governance is rendered problematic by the fact that people associate in groups that include only parts of society. And as intersubjective experience is locally socialized, the perceptions, habits of thought, values, attitudes, and beliefs of particular groups diverge. Put another way, intersubjective experience both unites and divides. The intersubjective experiences we share with some set us off from others, and localized socialization renders incommensurate the expectations, perceptions, and understandings of the experiences different groups have. Moreover, we often begin to rely upon such divisions to secure our immediate interests, our identities, and whatever power we are able to muster. The resultant inevitability of localized intersubjectivities implies that intersubjective pluralism will in some cases give rise to irreconcilable differences between groups. In this sense, the values, attitudes, and beliefs that groups espouse are often incommensurable, frequently entailing conflict and oppositional relationships. Thus, what many have said about democracy is actually true of all societies: there is always "an active tension between cultural drives to identity [broad socialization processes] and the persistent ethical need to contest the dogmatization of hegemonic, relational identities."[51]

Moreover, there are limits to our acceptance of any particular group's authority; there are "boundaries to what has been called the zone of acceptance," and we look to or construct other groups to take up authority outside another group's zone.[52] Thus the pluralism of intersubjectivities multiplies; and this multiplicity is cross-cutting when any individual plays multiple roles in society, necessitating membership in an assortment of groups. Cross-cutting, in turn, destabilizes particular groups when individual members introduce ideas, analogies, metaphors, perceptions, values, and beliefs from outside groups. Given this dynamic, the views, values, and allocations of power within groups are not fixed but contingent, shifting in ways that cannot be prescribed. They need to acknowledge the dimension of power and antagonism and their ineradicable character.

For these reasons, coherent agreement within a group is as rare as an overarching society-wide consensus. While elements of consensus may exist in a group or society, they rarely prevail and never seem to remain fixed. Consequently, it is probably erroneous to think that some kind of collective voice can emerge from disparate groups, and neither administrative theory nor administrative practice can rely upon such voices (any more than they can rely upon "identity thinking") to direct their endeavors toward attaining the broadest possible intersubjective experience of good governance.

Fortunately, the intersubjective experience of good governance does not require the identity of particular or broadly shared views or values with those of administrative agencies. There are two reasons for this. First, the conditions giving rise to this experience are diverse. They might include, for example, the uncritical acceptance of "broadly socialized" values, pragmatic considerations, ideological or philosophical commitments and advantage. Consequently, there is simply no need for a single, all-encompassing identity of interests, values, and beliefs so long as a broad, underlying, intersubjective experience of good governance is maintained for whatever plurality of intersubjective reasons. Second, the intersubjective experience of good governance is neither an indiscriminate nor an all-or-none affair. The experience admits of gradations regarding both the extent to which administrative institutions (generally or particularly) achieve good governance and the relative contribution of different institutions to the maintenance (or failing) of good governance. For these two reasons we might say that there are zones of accepted deviation from both local and broadly shared intersubjectivities that are defined intersubjectively (broadly and locally) for both different institutions and institutions as a whole. Consequently, administrative theory and practice need only direct their endeavors toward attaining the intersubjective experience of good governance within the zones of acceptable deviation. Of course, these zones will change with experience, and administrative agencies must monitor these changes and evolve themselves accordingly.

The Problem of Agency

Another problem entailed by our argument regarding the nature of public administration's subject matter involves the question of human agency. Unless humans are capable of shaping their circumstances and thus avoiding subjugation to both nature's regularities and tradition's dictates, it makes little sense to take up the intersubjective experience of good governance as a subject matter. In fact, without a capacity for agency, it makes no sense for people to seek or to understand anything at all, except perhaps out of idle curiosity. As largely dependent variables, there would be little if anything we might do about our situation beyond what is necessitated by the forces acting upon us.

Now, contrary to our intersubjective experience of agency, many scholars hold that all human action is in fact determined. We find among the ancient Greeks and Romans, for example, the conviction that our entire lives are predetermined by an inexorable fate or destiny. Similarly, the central tenets of many religions assert the omnipotence of one or another god, and from this many conclude that everything is determined by the will of that god. Even where such beliefs posit free will as a co-tenet, the idea remains that divine will cannot be thwarted and that undertakings contrary to the divine plan shall inevitably be set right. Materialists aver that the religious are correct about this, but for all the wrong reasons. They admit the existence of nothing but "matter" and ascribe all behavior to the properties of that matter acting and reacting according to discoverable chemical, physical, and mechanical "laws."[53] To materialists, what we perceive as our agency is actually determined by physical factors (including cerebral functions, motor impulses, and kinesthetic sensations) grounded in our hereditary constitution and our corporeal environment; and modern psychology concurs, holding that the will is not intrinsically free but is determined by antecedent psychical as well as physical conditions and causes. The only loosening of this rigid cause-and-effect necessity that is countenanced by determinists is the suggestion that the properties of some particular matter may be contingently related to causes and effects in certain circumstances—that is, the properties of a particular matter may be such that (1) it does not always react, or (2) it does not always react in the same way, or (3) it may or may not always produce the same effect. Hence, causal "laws" may be statistical, and noted effects may be "propensities" or "statistical tendencies."[54] Nevertheless, there is in all cases a finite range of possible results to which our behavior is constrained, and all possible events remain predictable.

Sociological determinism is added as a sort of nuance to physical, chemical, mechanical, hereditary, psychical, and statistical determinism. This form of determinism holds that as broad socialization into shared concepts and epistemologies is necessary to both meaningful thought and meaningful communication at both the broad and the local levels, our meaningful, successful participation in any society requires that we think and talk in the terms constructed by that society. Hence, we are always dependent variables acting and reacting according to some society's constructs. For example, one variant of this logic holds that if we ever were agents, we forfeited our volition piecemeal through the historical construction of rationally administered consumer societies. In brief, the argument suggests that the technorational desideratum of consistency inherent to any rationalized consumer-oriented society requires that it be administered by public institutions erected upon the assumption that citizens are utility maximizers functioning within a technorational consumerist system.[55] Successful navigation through such a system requires that individuals not only think technorationally and unambivalently in means-ends categories, but trust their fate to large-scale organizations that administer the apparatus that is rationally and scientifically constructed to pursue rational consumerist ends. Thus, bureaucratic power, legitimated through dominant discourses, subjugates individuals subtly but nevertheless aggressively until they are denied both their true interests and their ability to think in any other way. In the end, the logic of this approach returns us to the conclusion that we never were agents because the concept of individuals as free, reflective, creative, and self-determining agents is necessarily another construct invented by the dominant technorational ideology and discourses embedded in public agencies.[56]

For these reasons, determinists of all stripes assert that our intersubjective experience of agency is a mere illusion that arises from our ignorance of the causes necessitating our behavior. Theoretically, this conclusion seems to be grounded in the assumption that because people are composed of familiar matter sporting known and well-understood properties and processes, every state, reaction, and effect of the human being must be predictable. Methodologically, the idea seems to be that if volitional acts can be analytically reduced to physical, chemical, mechanical, psychical or sociological causes, then any theory of agency must collapse.

Dissatisfied with the determinist appeal to ignorance, we contend that this thesis is in error. First, we should point out that determinists labor under ignorance as well; they do not know that our voluntary acts are necessitated any more than we know that they are freely chosen. We simply have stronger reasons for concluding that human agency is the case. Briefly, we have intersubjective experiences of both volition and determination, and these experiences are as real (as nonillusionary) as our intersubjective experiences of matter, psychical causes, and physical processes. Thus the conclusion (theory) of human agency is not only as ontologically grounded as the theory of determinism, it accounts for more as well. Determinism, that is, simply denies the reality of experiences that do not fit its worldview—that is, the view that "all of the surface features of the world are entirely caused by and realized in systems of micro-elements, [and that] the behavior of micro-elements is sufficient to determine everything that happens."[57] Agency theory, however, confirms both sorts of experience intersubjectively, accepts their equal ontological status, and concludes that people are so ontologically constituted as to enjoy the properties of both volition and determination. It does not deny that we are familiar matter, but only that such matter in its human configuration is incapable of either reacting to circumstance in anything but predeterminable ways or of producing anything but predeterminable effects. That is, given that particular organizations of matter often result in novel emergent properties (e.g., the "mind," the determinist's "psychical causes"), "if a certain configuration of matter is all that is required, to make it think rationally, it will be impossible to show any good reason why the same configuration may not make it act rationally and freely."[58] Hence, because agency theory is ontologically grounded and does a better job of explaining intersubjective experience, it is a better theory than physical, chemical, mechanical, hereditary, psychical, and statistical determinism.

What of sociological determinism? Given the argument so far, there seems to be no general problem of agency at the sociological level either. That is, as only ontologically constituted beings can be rendered cognitively self-conscious,[59] the fact that the attitudes, values, and beliefs by which we define ourselves are initially derived from contact with others does not preclude agency. There are two reasons for this. First, as ontologically constituted beings, we constitute our "selves," "in an active fashion, by the practices of the self, [even though] these practices are . . . not something that the individual invents by himself, [but] are patterns that he finds in his culture and which are proposed, suggested and imposed on him by his culture, his society and his social group."[60] Consequently, socialization and the learning of rules and techniques for operating pragmatically within our historical, cultural, and socioeconomic context do not result in "the identification of the self with others,"[61] but in "the growth of [our] own understanding of [ourselves] . . . pursued through [our] understanding of the other."[62] Of course, problems of agency do arise sociologically in those local situations where the recognition of concepts and epistemologies as discardable tools is absent. In such situations, the problem arises, how can we accomplish the necessary critical distance from the very intersubjectivity that defines us initially and renders us self-conscious?

Second, our situation as selves embedded in a particular immediate lifeworld from which we constitute our personal identities, habits of thought, and particular ways of seeing is regularly disturbed by the habits of thought, practices, and behaviors of those embedded in lifeworlds of their own. Although others may share a socialization process with us, the effects of that process on each of us is conditioned by the immediate practical necessities of our individual socioeconomic and historical circumstances. Consequently, the individual identities and perceptions that we each actively constitute experience a host of individual perspectives offering alternative conceptualizations, attitudes, values, and beliefs, and it is these differences that provide the opportunity for us to exercise our ontologically constituted capacity for agency.

The Problem of Nihilism

Because intersubjective experience is historically, socially, economically, and institutionally contingent, all categories for cognition, all systems of thought and discourse (normative and empirical), all methodologies, and all institutions are socially and historically constructed, controlled, and conformed.[63] From this, many conclude that we are adrift without normative, empirical, or methodological stars. This situation actually leads to as much hope as dismay.[64] Hope arises from the fact that "even small fluctuations [in our intersubjective experience] may grow and change the overall structure" of our lives.[65] Consequently, "individual [experience and] activity is not doomed to insignificance."[66] On the other hand, there is dismay because the security of stable, permanent rules, structures, and habits of thought seems gone forever[67] Unfortunately, many find little consolation in the fact that the smallest fluctuations in the way we see and understand can change all things, and for them it has proven a short step to radical relativism, pessimism, and nihilism. If, they argue, all concepts of justice, equality, emancipation, tolerance, and happiness are grounded in nothing other than contingency, they hardly seem reliable guides to "good" practice. In fact, they continue, the mere idea that justice and freedom are better than injustice and oppression gains no purchase. This nihilistic form of relativism is an important problem for both theory and practice in public administration because it suggests that normative and empirical theories are nothing more than embodiments of contextual and cultural influences. Consequently, they are not only pointless but potentially harmful, and public administration as a social science, or a normative discourse, or a field of practice is at the very least a profoundly irrational endeavor.

Of course, this position is paradoxical because it is in fact just another form of theory[68] and therefore must, itself, be contingent and not universally applicable. But more important, this sort of nihilism follows only from a disappointment with the fact that the grounds for a particular kind of certainty (i.e., the grounds posited by teleological and positivist presuppositions) have failed analytically and empirically. However, what many of the disappointed perhaps overlook is that this failure of teleological and positivist groundings for our concepts, categories, norms, discourses, methodologies, and institutions renders them neither unacceptable nor incoherent nor merely conventional. It merely means that we must look for our groundings and guiding stars on earth rather than in heaven, and among ourselves rather than in the givens of nature.

For example, all that we have constructed is not unacceptable because it lacks teleological or positivist groundings. It is true, of course, that because the values, attitudes, theories, and traditions by which we live are constructed by us and operate only within the dynamic context of social life from within which they arise and within which they have their application, they cannot be successfully turned around to depict, explain, or direct the dynamics themselves. All such attempts result in either circularity (as, e.g., when living according to Christian values and traditions is explained by its being biblically enjoined, and attending to the Bible is explained by the fact that one is Christian) or infinite regress (as, e.g., when one explains living according to Christian values and traditions as being biblically enjoined, explains attending to the Bible as being required by God, explains attending to God as what we should do, explains doing what we should as proper behavior, etc., etc., etc.). However, this is not cause for despair. Once we realize that "giving grounds . . . [at some point] comes to an end . . . [and that] at the end of reasons comes persuasion,"[69] we realize that we do not require teleological or positivist groundings for accepting all that we have constructed; we only need agreement among ourselves. It is only in the absence of such agreement that our concepts, norms, and institutions are unacceptable. At that point, "[we] cannot be reconciled with one another . . . [and] each man declares the other a fool and an heretic."[70]

Nor is it the case that we require normative or positivist groundings for all of our concepts, norms, and institutions to be coherent. The meaning of a good society, for example, is not given coherence teleologically, or by nature, or by our traditions, attitudes, and values alone. Its coherence depends ultimately upon our intersubjective experiences as we live out the ensemble of practices encountered in both our lifeworlds and our societal roles and constituting our form of life. As those experiences change, so does the nature (the meaning, the "essence") of a good society and what we must do as agents to secure it. The meaning or nature of a good society is in this sense indeterminate (incoherent), but while its meaning is a matter of neither strict logic nor practical necessity, it is not incoherent or arbitrary either. It is socially and humanly coherent and sensible. It is not essentially any particular thing, but it can not be just anything at all. The meaning of a good society is coherent exactly to the extent that our form of life and its transformations are coherent.

Moreover, the norms we establish through living out the dynamics of our social lives are by no means merely conventional. While they are not transcendental or external norms, revealed, discovered, or abstractly derived as independent criteria, they do constitute sociopsychological constraints, acquired from both training and social practices, but (at least initially) developed intersubjectively as practical responses to our intersubjective experience. Because they derive from experience in practice, they are as dynamic as our form of life and our language games. But even though they are not fixed or static, and while they are discardable when they no longer serve us well in practice, they are not discardable wholesale or simply at our whim. They constitute the social grounding and understood meanings upon which we must build, or with respect to which we must distinguish and justify new norms and practices. For example, when we try to determine whether a particular choice accords with our previous successful practice or is perhaps worse or something new and better, the determination of what counts as an accord, or what counts as a successful adaptation of a practice to a situation, or what counts as something different can be answered only within the ongoing social practices and norms. There is just no place else to look for reference to act. Consequently, these practices are not merely conventional because they cannot be lightly discarded nor discarded wholesale, but must be changed intersubjectively from within ongoing social practice.

Core Concerns

By claiming ontological status for public administration's subject matter, we are suggesting that so long as public administration's theory and practice act upon the cognition of that subject matter (i.e., act upon what they have come to know about how the intersubjective experience of good government is secured and maintained) and do not act regardless of it or contrary to public administration. Public administration has a legitimate claim to both the exercise of power and the judgment that it is acting justly as well. Of course, as institutionalists suggest, administrators may propagandize and manipulate in an attempt to produce a widely shared intersubjective experience of good governance. But for the reasons delineated above, it is impossible for these attempts to have lasting success or to have a totalizing impact. Hence, to an important extent they will fail. To be successful, public administration's theory and practice must therefore take as core concerns those intersubjective experiences that account for this failure and recurrently thwart intersubjective experiences of good governance.

What are these core concerns? As argued above, the intersubjective experience of good governance is a de facto matter, and it is based not upon the identity of particular or broadly shared experiences, views, or values with those of administrative agencies, but upon a broad, underlying intersubjective experience of good governance that may be obtained for a plurality of intersubjective reasons at the local level and may be broadly maintained so long as administrative agencies act within zones of acceptable deviation. At a minimum, it seems clear that the requisite intersubjective experience of good governance cannot be secured or maintained within broad and localized zones of acceptable deviation, unless public administration theory and practice concern themselves with issues of power, emancipation, and social justice. This is because the experience of good governance is negated and institutional legitimacy is diminished when institutions interfere unnecessarily or too extensively with the ability of people to meet their interests or when people experience inequitable treatment, marginalization, suppression, or degradation at the hands of institutions.

In practice, these recurrent obstacles to the intersubjective experience of good governance suggest that the character of administrative institutions (i.e., their formal and informal structures, processes, and cultures) must be such as to counter intersubjective experiences of powerlessness and inequity while fostering intersubjective experiences of liberty. These requisites, in turn, constitute core concerns for theory that must be addressed if its claim to cognitive authority regarding its subject matter is to be honored. As argued above, it is our contention that these core concerns can be dealt with most effectively by a critical theory informed by an evolutionary epistemology that is built upon a worldview informed by a synergy among institutional, behavioral, and hermeneutic methodologies. This contention will be argued in detail in subsequent chapters, and toward that end we will briefly outline what are, for our purposes, the important implications of these core concerns for theory and practice in public administration.

Administrative Institutions, Theory, Practice, and Power

"A claim to cognitive authority [i.e., a claim to know that something is really the case] is almost inseparable from a claim to practical authority."[71] It is, more particularly, a claim to the legitimate exercise of power, especially the power to direct behavior along "proper" channels. Thus, if public administration and administrative institutions are to claim power (practical authority) within both the academic and public spheres, they must ensure that they remain grounded ontologically by closing gaps between theory, practice, and the intersubjective experience of good governance. Otherwise, such gaps leave practice to the vagaries of "accident and force"[72] and theory to the vagaries of bias and arbitrary power. Thus, for example, we may find ourselves experiencing either "hyperliberalism,"[73] wherein judgment is immobilized by conflicting individual experiences, or authoritarianism, wherein bureaucrats and "experts" force judgment into strict conformity with their singular theories.[74]

With regard to the intersubjective experience of powerlessness, then, public administration theory and practice must remain aware that any broadly shared intersubjective experience exists only temporarily and always through many different and conflicting interpretations. Consequently, they must accept "discensus," conflict, and regularly recurring experiences of powerlessness as inevitable, and attend to developing and maintaining institutional structures, processes, and cultures capable of discovering and closing the theory-practice gap that can lead to intersubjective experiences of powerlessness as they develop. At an absolute minimum, of course, this suggests that the character of administrative institutions must be such that no local or broad intersubjectivity can claim to represent the totality and thus claim absolute authority. But more important, it suggests that the main endeavor of administrative theory and practice should not be the elimination or regulation of power or opposition (e.g., through consensus building, compromise, scientific findings, or marginalization—thereby masking both irreconcilable differences and conflict), but the dedication of a public place where people might develop and confirm the intersubjective experience of empowerment and hence good governance. To advance the intersubjective experience of good governance, then, administrative institutions must provide formal and informal venues that enable diverse voices to be heard in a variety of contexts and regularly over time. Thus administrative institutions must view themselves as among those "medium[s] through which the absence of consensus is interrogated."[75]

In other words, as many have pointed out, the multiplicity of intersubjectivities and the necessity for a shared form of life (broad socialization) to produce the ability to think and evaluate meaningfully requires a multiplication of spaces where the widest variety of intersubjectivities may be heard effectively by the greatest number, and a willingness by everyone to hear what everyone else is saying.[76] Not all people will be able or want to participate in electoral politics, interest groups, or administrative institutions that make it difficult for their voices to be heard. It is therefore necessary to develop administrative institutions in which local intersubjectivities can be heard and to foster the intersubjective recognition that administrative agencies contribute to this process as public spaces.

Administrative Institutions, Theory, Practice, and Social Justice

When public administration theory and practice claim to act upon cognitive authority, they claim to be exercising power not only legitimately but justly as well. To retain cognitive authority in normative matters (and hence to be acting "justly"), administrative theory and practice must therefore ground their norms in intersubjective experience, which must provide the justification for critiquing and revising extant norms, institutions, and practices.

Now, we experience injustice when "others harm us in ways that we expect them to avoid."[77] The problem is that there is a certain amount of inertia inherent in our experiences of injustice, and this inertia is often experienced in our relationships with administrative institutions. That is, because experiences of injustice involve expectations, they involve an a priori just order that is already in our thoughts and that is violated when people act in certain ways.[78] As all things change, however, there will be changes in our expectations and experience of harm. When these changes occur, the affective dimension of intersubjective experience (the feeling of being wronged) conflicts with the objective conclusion that the right thing is being done. Thus, even when others behave in ways that do not violate the apriority, shared feelings of inequity or mistreatment arise and begin to disturb the intersubjective experience of good governance even though we may not feel justified in having them because they cannot be easily articulated in justificatory terms recognized by the apriority. Institutional insistence upon an a priori standard under such changed circumstances, then, amounts ultimately to an attempt at subordinating our intersubjective experiences by limiting our expectations of one another and training us to regard only certain kinds of behavior as unjust.[79] To the extent that this attempt is successful, inertia results.

Overcoming this inertia requires a process of critique that involves a "deconstruction" and reevaluation of concepts, categories, and understandings currently taken to be natural or objective, and often requires restructuring the institutional anchors of these concepts and categories as well. In addition, it requires a positive critique pointing the way toward the construction of an alternative intersubjective set of concepts and categories upon which new expectations might be based. Administrative theory and practice, therefore, must develop in such a way as to actuate these negative and positive critiques with regard to institutional arrangements, the capacity of these arrangements to meet the needs of community members, and the various empirical conditions that bear on these matters. Moreover, to be meaningful and effective, these critiques must also be capable of reformulating traditional normative concepts and categories to better address existing social conditions, needs, and possibilities for change. Finally, those methods of critiquing must include in their foundational assumptions the realization that any emergent intersubjective experience of just treatment is a moment in ongoing social experience that arises within a historical and cultural context that will change. Hence, they must not equate justice with any single essence, as many have done, for example, by equating justice with different forms of equality. Whether equality as justice is ontologically grounded depends upon current and historical intersubjective experience, and while an ontologically grounded ethos of equality might be thrown up from intersubjectively shared individual experiences and then become embedded in practice, discourse, and institutional structure, it may not.

Administrative Institutions, Theory, Practice, and Emancipation

Public administration in both theory and practice wrestles relentlessly with the notion that government regulation is always to some extent undesirable because it intrudes upon individual and group autonomy; and it is presupposed that individuals and groups (at least under normal circumstances) positively prefer liberty and autonomy, and that government regulation through administrative institutions is presumptively undesirable because it conflicts with that preference. These presumptions arise predominately from the idea that people possess certain "rights" as a matter of natural law, and that "liberty" (the right to possess one's own body and to determine one's own course of action) is among these natural rights.[80] Accordingly, it is thought that in order to garner the advantages of social living, people must exchange liberty for order and safety, the only question being how much liberty people need to relinquish. Some hold that we need relinquish most of our liberty (retaining necessarily the right of self-preservation),[81] while others insist that we must retain almost all of our liberty subject only to the state's providing a very good reason for limiting it and then only to the minimally necessary extent.[82] In opposition to this general idea that our natural liberty is necessarily forfeit to some degree when we construct a society, some suggest that as human beings our true "natural liberty" is not the liberty we experience in nature (which is in fact oppressive), but is to be found only in society (i.e., "natural liberty" for human beings is "civil liberty").[83] In the view of all of these "natural rights" philosophers, however, the state and its administrative institutions remain at least a threat to liberty as they may remove it altogether or thwart the general will.

The problem with this line of thought is that, as argued above, our concepts, categories, and epistemologies (including our concept of liberty and the natural rights epistemology within which it operates) are socially constructed; they are the result of broad and local socialization processes that are themselves the product of historical and cultural process intertwined with the development of the state and its institutions.[84] In other words, our intersubjective experience of liberty is to a significant extent dependent upon rather than necessarily in contradistinction to our institutions and practices. Put another way, society and its institutions are formed out of the recognition of mutual dependence and the perception of a possible future that can be built only upon a shared past. Our liberty (e.g., our emancipation from the dictates of nature), then, is partially the result of institutional processes that enable us to actualize our agency (our freedom to choose). Thus, the argument that administrative institutions and liberty are necessarily in opposition loses much of its force. Thus, meaningful administrative theory must remain cognizant that administrative practice is as necessary to liberty as it is threatening to it. We have established our institutions to free us as much as possible from, for example, the ravages of natural disaster, disease, and the incapacities of old age. We also establish institutions to free us from oppression by others (e.g., interpersonal violence, the power of concentrated capital and organized interests).

Conclusion

The arguments presented in this chapter are not completely incompatible with public administration's understanding of its disciplinary matrix or the commonly held understanding of how the field operates both in theory and praxis. There are people who will remain in opposition to such an idea, claiming that public administration is not a discipline nor can it have ontological status. We believe that though our argument may not be definitive on this topic, it is sufficiently developed to allow us to make the claim that public administration has both a disciplinary matrix and ontological status. In addition, we have demonstrated that these two aspects of public administration are inextricably linked, moving us away from many of the criticisms in this area toward more fruitful areas of study.

Although having both a disciplinary matrix and ontological status is useful both in theory and praxis, we must move our argument substantially further to truly apprehend the nature of public administration. Our next step must enable us to determine if public administration is or even can be critical in nature. If we determine it is or can be effectively grounded in critical theory, we must then understand what limitations such a linkage brings. The potential limitations associated with adopting a critical approach arguably can be mitigated or corrected by approaching critical theory from an evolutionary perspective.[85] Once we establish that public administration is both evolutionary and critical, we can conceive of a discipline that is truly able to grasp the complexity and depth of theory and praxis.

It becomes imperative that we first define, explain, and understand the nature of critical theory. In addition, we must then develop the notion of an evolutionary theory, thus enabling us to wrestle with the fundamental concepts at the core of theory and praxis. Once these are established and clarified, we can then fully grasp the nature of debates in public administration and arguably diffuse or conclude a number of them by framing them in an evolutionary critical theory context, allowing us to fully apprehend the complexity of each argument.

4

Critical Theory and Public Administration

The issues of how we arrive at the intersubjectivity of particular particulars (objects and patterns), and how we arrive at what particular particulars "mean," lie at the heart of what constitutes good (rational and practical) empirical and normative theory in public administration. How these events transpire, for example, conditions strongly both the essential elements of a good empirical theory (i.e., how its concepts, categories, principles, and methods are formed) and the essential processes used in theorizing (e.g., the usefulness of and interrelationships among various methodologies and inductive, deductive, synthetic, analytic, critical, and creative thinking). Of course, how this intersubjectivity of meaning and particular particulars occurs also conditions the content of our normative thinking. Concepts of "good" and "bad," for example, arise as responses in context and with regard to objects and patterns forefronted in that context. Moreover, both the source of such meanings and particular particulars and the way that they come about suggest which sets of practices, institutions, and values should be advanced. For example, which of our concepts, categories, practices, institutions, and values arose in an ontologically grounded way (i.e., from the intersubjectivity of a shared subject matter as opposed to the particular particulars of an ideology)? Those that did certainly constitute an articulation of an intersubjective experience that is "not mere domination; like the traditional authority embedded in historical values and hierarchically structured power relations, but the product of communal negotiation of needs in response to the threats of nature."[1] Being grounded in reality, such practices and institutions are probably better than not being grounded at all. On the other hand, institutions and practices grounded in ideology are probably worse, whether that ideology is one of power and privilege or equality (it may be that privilege is negotiated communally and that equality is not ontologically grounded under the exigencies of certain contexts).

As argued above, in order for theory in public administration to remain ontologically grounded and so produce rational and practical normative and empirical theory, its methodology must develop a synergy among institutional, behavioral, and hermeneutic approaches and methodologies. That is, in order to forefront truly relevant patterns and particular particulars given the nature of its subject matter and the practical focus of the discipline and practice, public administration theory must devise a way of making institutional, behavioral, and hermeneutic approaches work in concert (in an interactive and mutually developing way) toward not just a multidimensional but a holistic understanding of its subject matter. Only by doing so might it shape the results of its inquiry into dynamic explanatory models capturing the synergy among behavior, meaning, values, and institutional factors as they affect the flux and evolution of intersubjective experience and institutional change.

It is our contention that the methods and techniques developed by critical theory can accomplish the necessary synergy when employed from within an endogenous evolutionary paradigm. To establish this contention, we will first explain in greater detail why a synergy among traditional social science, institutionalism, and hermeneutics is advantageous to public administration, and why each approach alone is either inappropriate to public administration's subject matter or insufficient to its theoretical requirements. We will then delineate how critical theory develops and maintains this synergy by keeping scholars and practitioners conscious that objects, patterns, and relationships can always be forefronted and analyzed through multiple conceptual, normative, theoretical, and historical perspectives, and by encouraging multiple constructions and analyses through ongoing critique. In the next chapter, we will consider certain telling criticisms that reveal why critical theory as traditionally pursued (i.e., as pursued from within a radically democratic paradigm) cannot fulfill this role. Finally, we will explain how critical theory as employed from within an endogenous evolutionary paradigm does not suffer from these criticisms and hence works to develop both the necessary synergy and the sort of practical and rational theory public administration requires.

The Useful Insufficiency of Traditional Social Science, Institutionalism, and Hermeneutics

Traditionally, scientific social research, institutionalism, hermeneutics, and critical theory focus on very different things in order to describe and explain social phenomena and social behavior. Social scientific research traditionally seeks empirical evidence about how most people behave. Its primary goal is to develop the broadest possible and most parsimonious explanations for the behavior of large collectivities and to employ such covering theories to evaluate, critique, and suggest effective ways for dealing with both individual and collective behavior in specific situations. Institutionalist research, in contrast, traditionally utilizes "anthropological, historical, demographic, political, and psychological data"[2] to model the unique patterns of social evolution displayed by different societies as they are each driven by technology, guided by human agency (instrumental valuing), advanced or retarded by institutions, and conditioned by power, conflict, and vested interests.[3] The goal is to "uncover the unique character of the specific [evolutionary process] under investigation, [and] to discover its unique patterns or regularities."[4] Hermeneutics differs from both in analyzing language and symbol use (and nonuse) as it is contextualized socioculturally in discursive traditions. Its goal is to uncover and analyze meaning in everyday life, to expose hidden meanings, and to reconstruct original intent.[5] Critical theory, in contradistinction to all three, traditionally employs qualitative and analytical techniques such as concept analysis, deconstruction, case studies, immanent critique, action research, and hermeneutics to study culture and how cultural processes develop and maintain meanings, attitudes, values, beliefs, epistemologies, and power relationships among the members and institutions of a society. The primary goal is to reveal the impact of culturally accepted ideologies, patterns, and meanings on individual and group lifeworlds and psyches.

All four endeavors hope to empower, to make available the means of improving upon primal realities and to give people some control over both the collective and the individual human situation. Traditional social science hopes to do this by providing us with the general laws of social behavior and social phenomena from which we might deduce the proper actions to take in pursuit of our ends. Institutionalism seeks to deepen our understanding and so empower us by revealing the tendencies and implications of regularities and patterns found to be actually operating in different concrete social realities. Hermeneutics seeks to deepen still further and complete our understanding by revealing the meaning and import of social phenomena, patterns, regularities, and behavior to those actually involved. And critical theory seeks to broaden our understanding and to point beyond what is the case by bringing to light the gaps between what a society's ideology promises and what it delivers (between what it intends and what it does), and by suggesting ways of empowering people to act as agents in pursuit of their interests by removing any unnecessary constraints found in the regularities, patterns, and meanings that are imposed by culture, discourse, history, and institutional character.

Public administration seeks the same goal by putting study to practice at that point where the general meets the specific; where the social and collective meets the institutional, the group, and the individual; and where the behavioral, psychological, institutional, and historical-cultural conjoin. In this endeavor and because of its subject matter, public administration traditionally oscillates between traditional social scientific studies of collective administrative and institutional behavior; the clustering of incommensurable and incompatible theories, techniques, and methodologies; and the containment of the breadth of its study and practice to theories and methodologies bearing family resemblances to one another, as the academic, political, economic, and social situation requires. But always its thought and practice remain at that juncture where manipulation of the environment meets power, interest, and ideology; where objectivity meets humanism and self-reflection; and where objectification meets hermeneutics, culture, history, and belief. And at that point, its central goal is always the realization of the intersubjective experience of good governance.

This positioning and goal of public administration are clearly reflected in its research concerns, which, despite the oscillation among methodologies, have changed relatively little over time.[6] These concerns may be conceptualized in terms of authority (power struggles, politics, citizen demands, and resource allocation), legitimacy (the constitutional grounding of the field, the cultural and historical context of democratic values, public response to institutional behavior and values, and the law), choice (realistic options in context), and decision making (keeping the government functioning, effectively and ethically).[7] Thus, public administration research is always aimed at determining how institutional structure and process fit to culture, context, (political, social, and economic) and citizen expectation to secure the intersubjective experience of good governance; and to the extent that administrative theory and practice are detached from intersubjective experience, public administration's scholars and practitioners lose their rational (practical) authority.

Public Administration Theory and Traditional Social Science

Of greatest concern to public administration, then, is the possible detachment of administrative theory and institutional practice from the reality of intersubjective experience. Maintaining this attachment cannot be accomplished through the methods and problem-solving approaches offered by traditional social science alone, because that tradition understands social reality as a phenomenon generated according to discoverable laws about the inherent characteristics and behavior of the parts constituting a society, given a specified set of interrelationships among those parts. It thus does not take a thoroughly institutional perspective, and it is by design abstract, normatively disengaged, and constructed to be a useful tool in understanding social phenomena and patterns of behavior only broadly, generally, and objectively.[8]

For the same reasons, traditional social science theorizing, however well it might lead to an understanding of the common good and so provide certain insights for critiquing the way things are, lacks the "mutual deliberation necessary to achieve and sustain it."[9] As many have pointed out, disengaged ivory tower or scientific philosopher kings are like the Lady of Shalott, dealing only with the images of things and condemned to finding all they see unsatisfactory.[10] Moreover, traditional scientific theorizing is inappropriate to public administration's subject matter. These theories take their subject matter to be relatively more fixed ontologically than is the intersubjective experience of good governance. In other words, traditional social science as a whole cannot offer the necessary sort of methodology or ongoing critique necessary to adapt institutional character so as to secure and maintain the intersubjective experience of good governance.

Nevertheless, traditional social scientific methods are still useful. This is because the intersubjective experience of good governance is a de facto matter that may be obtained for a plurality of intersubjective reasons and may be maintained so long as administrative agencies act within "zones of acceptable deviation." Consequently, while its existence and composition are sufficiently variable in time and place to render parsimonious transtemporal or transspatial covering laws problematic, we may expect to find underlying regularities in, for example, the interaction of political, social, and economic variables that affect the intersubjective experience of good governance among those sharing a form of life (i.e., certain agreed-upon presuppositions, common habits of thought, and conventional patterns of behavior and discourse), and perhaps to formulate those regularities mathematically. In other words, as coherent behavior occurs at particular times and places as a result of intersubjective experience, it can be studied using traditional social scientific methods even though the subject matter under study is ontologically fluid rather than fixed. Consequently, in order to employ positivist methods fruitfully, we do not need positivist or transcendental groundings to understand how the intersubjective experience of good governance might occur at a given place and might be maintained and transformed coherently over time.

Public Administration Theory and Institutionalism

Institutionalism offers certain advantages to public administration over traditional social science in that it is holistic and evolutionary. That is, institutionalists see social reality as an evolving and somewhat unpredictable process that is directed both by the inherent characteristics of a society's institutionalized relationships and by a limited human agency that is definitively restricted and directed by the society and institutions from within which it arises.[11] Thus, institutionalism is founded upon

> an underlying presumption of the "radical indeterminacy of reality." The world is in constant flux. There is little reason to believe that the regularities found in one socio-economic context will be found in others. And although this skepticism may be more strongly held by the theory agnostics, even those institutionalists who are active theory seekers appear to adhere to a degree of relativism concerning the subject matter of economics that far exceeds that held by neoclassical, Keynesian, or even orthodox Marxist economics.[12]

That is, socioeconomic reality is not seen as a phenomenon arising and stabilizing primarily according to discoverable "laws" about the nature and behavior of the "parts" constituting a society. Rather, it is a phenomenon (1) that is continuously evolving from the relations both among the parts and between the parts (including a limited human agency) and a coherent whole, and (2) that can be understood only in terms of the whole.

Because institutionalism takes its subject matter to be a holistic, evolutionary, and somewhat unpredictable process, its methodologies (e.g., participant observation, case study, pattern modeling) are all geared to understanding social evolution in situ, and its theory does not include "sets of formal relationships that might permit axiomatic modeling."[13] Moreover, institutionalists view this holistic evolutionary process from a wholly institutional perspective. That is, as opposed to traditional social scientists, who adopt a behavioral perspective by taking human beings as their unit of analysis, "institutionalists hold the view that all socially relevant behavior is learned [from institutions] and is, for the most part, habitual."[14] Consequently, "society" is an "institutional system" or "set of institutions"[15] that "socialize individuals by creating habits as to how they should behave in a prescribed manner."[16] And the primary forces for social change are neither external circumstances modifying human behavior nor technological change per se, but the intersubjective experience ("a sensed awareness")[17] "within the community [of] a need to modify habitual patterns of thought and behavior;" arising from "frontier experiences" that break down allegiances to traditional patterns of behavior; "environmental catastrophes"; and "contact with other cultures through war trade."[18]

All of this is particularly useful to public administration's theory and practice, given its subject matter and its locus of study. As explained above, the intersubjective experience of good governance is by its nature evolutionary, de facto, tinged with unpredictability due to human agency, and relative to the broad and local socialization processes operative in particular societies; and it is this intersubjective experience that is the motive force behind change. Consequently, public administration would benefit by employing many of institutionalism's methods and would find many of its theoretical insights helpful. Similarly, while the unit of analysis in public administration is "people" rather than the individual or the society's institutions, it is people as institutionally affected (i.e., as affected by social rules, norms, roles, and "the settled habits of thought common to the generality of men" that guide behavior.[19] Therefore, institutionalism's institutional perspective is informative and helpful as well.

However, an understanding of how to maintain the attachment among theory, practice, and the reality of intersubjective experience cannot be attained through the use of institutionalism as a whole. For example, while institutionalists "have recognized that formal methods—whether of the a priori rationalist type or of the logical positivist covering law model—fail to explain the nature of social reality,"[20] they nevertheless explore patterns of relationships among the parts and between the parts and the whole of the social system from an objectivist point of view. This has three undesirable consequences so far as theory in public administration is concerned. First, such an approach does not provide the synergy public administration theory seeks in a methodology. Rather it eschews traditional social science by denying its efficacy, and it eschews hermeneutics by implying that a sufficient understanding of social behavior and phenomena can be derived from "observable arrangements of people's affairs . . . [rather than from] characterizations of people's activities deriving from assumptions, intuitions, or introspection."[21] Second, its objectivist orientation implies an assumption that we can experience reality rather than just having experiences of it. Hence, it sees theory as a mental construct for organizing our perceptions of an ontologically independent, factual institutional reality that is observable directly through sensory experience. The problem with this for public administration is that, as explained above, institutional practices, behaviors, and discourses are neither ontologically nor experientially given but generated and known only through (intersubjective) conceptual systems, theories, and (often tacit) metatheories. Moreover, they are distilled to manageable proportions primarily through interpersonal discourse in a range of languages (e.g, scientific, religious, historical, legal) that are constructed according to particular interests and purposes arising from experience.[22] Thus, insititutionalists who focus upon these artifacts for theoretical purposes "confront a subject matter that is epistemologically and discursively (i.e., intersubjectively) pre-constituted and pre-interpreted"[23] by the individuals, groups, and institutions they study, and consequently not the proper subject matter of theory.

Finally, insofar as institutionalism takes a society's institutional structure as given, it is engaged ultimately in studying and prescribing for the elaboration of a particular society and culture through its particular institutions, and not in studying how that culture and its institutions might change in ways not implied by the present tendencies of its institutions, developing technology, and citizen agents as they are currently conditioned institutionally. Institutionalists thus proceed "by gaining a familiarity with the history of a problem, examining its legal foundations, and understanding its social and political aspects,"[24] in order to contribute to the better working of particular elements within the existing institutional structure, a structure which itself is taken as given. Hence they do not engage in the sort of critique that might suggest fundamental restructurings of institutions or radically different way of employing current technologies, should current structures and uses fail to secure or maintain the intersubjective experience of good governance. Rather they, "content themselves with the mundane assemblage of a 'mixture' of partial improvements."[25]

In addition to the problems entailed by its ontology, institutional theory does not allow itself to be instructed by its subject matter to the extent required by theory in public administration. Despite institutionalist exhortations to the effect that theorists should "put no more [theoretical] order in the facts of life . . . than actually exists,"[26] institutionalism impresses a certain teleology upon its subject matter, a teleology not of "end states" or "covering laws" but of certain dynamics that are assumed to be "always at work everywhere and under all circumstances."[27] In its simplest form, this teleology holds that technological change is the motor force of social change and that the technological impetus is "impeded and at times fully braked by (1) atavistic institutions," (2) the habits of thought and behavior that are created by institutions and socialized into human beings, and (3) vested interests. Normatively, institutionalism assumes that society is always and everywhere worse off because of these forces opposing the advances impelled by technology. Thus "correct" institutional values and choices are those that displace "ceremonial patters of behavior with instrumental patterns of behavior,"[28] and cooperation, as opposed to competition, is thought to further economic and social well-being.[29] Such a teleological overlay is inappropriate to public administration's subject matter. The intersubjective experience of good governance is highly contingent and the factors contributing to it (technological development being only one) vary among groups, from the broad to the local level, and within different zones of acceptance. Consequently, no singular dynamic that is "always at work everywhere and under all circumstances" can be anticipated. Moreover, as the intersubjective experience of good governance is de facto, no "always correct" values and choices will obtain.

A fourth problem revolves about the fact that institutionalism is overly deterministic regarding human agency. It holds, for example, that "human nature is a social or cultural phenomenon" and "not a biological one," because "man and society evolved together."[30] Now, what this means is not exactly clear. Does it mean, for example, that people are blank slates onto which their nature is written? If that is the case, what does it mean to say that "man and society evolved together"? Does this mean that there is a human nature, one that is "biological" but derivative of, or especially responsive to, or dependent in some other way upon social conditioning?

This ambivalence about human agency is a matter of contention throughout insitutionalist literature. On the one hand, for example, many institutionalists insist that "all behavior is cultural," clearly implying blank slate determinism[31] and relegating humans to the role of largely passive observers moving in obedience to laws operative within a culture and institutional structure that they can never control. On the other hand, some institutionalists promote "social value theory" or "instrumental value theory," according to which

> Human beings are culture-building creatures. In this they are unique for creative capabilities and represent a step beyond the limitations of organic or biological adaptation to the stresses of the environment in the direction of intellectual adaptation. Intellect and the human capacity for speech, the creation of language, and the creation of tools have allowed for selective and purposeful adaptation. Thus, humans adapt according to their own ideas and conceptualizations.[32]

This view clearly implies some degree of agency, suggesting that people make "value choices" among technologies and institutional options. However, this particular agency is restricted to choices among technologically and institutionally presented options and the values that these represent and inculcate into the individual. In the end, then, whether or not one is a "value theorist," "human nature becomes almost fully plastic such that few if any parameters are set on just how far humans are socializable";[33] hence there are few if any parameters set upon the extent to which human choices and values are determined by external forces.

As explained above, we feel that there are stronger reasons for public administration to accept the conclusion of a more extensive agency. Briefly, we have intersubjective experiences of our behavior being not only determined, but both constrained and unconstrained as well. These experiences are as real (as nonillusionary) as our intersubjective experiences of matter, psychical causes, physical processes, and institutional constraints. Rather than denying that these experiences are real or proclaiming that they are illusions, it makes more sense to accept human agency as ontologically constituted.

Public Administration Theory and Hermeneutics

Both traditional behavioral social science and institutionalism focus on describing, measuring, and explaining social behavior and social phenomena, leaving largely unexamined the intersubjective meanings, beliefs, and desires that constitute the reasons for the social behavior and social phenomena they study. Institutionalism takes this approach because it believes that meaning and belief are derivative of institutional character; traditional social science does likewise because it assumes that meaning and belief are revealed through behavior. Hermeneutics, conversely, focuses upon understanding the intersubjective meaning of such behavior and phenomena to those involved and affected.[34] Toward this end, it seeks out (1) the meanings expressed by signs and symbols (languages and texts) employed within a particular culture to deal with its historical experiences and current events,[35] and (2) the meanings embedded in a society's complex of normative, institutional, and behavioral traditions, because "the activity of a perceiving subject confronting a world is so radically identical with the activity of an interpreter confronting a text"[36] that such complexes are considered "susceptible of interpretation by hermeneutical methods."[37]

As our argument indicates, a theory and methodology that seek to reveal and understand whether the intersubjective experience of good governance is abroad in the land cannot reduce their study to behavioral or institutional correlates alone as they are likely to be misinterpreted without a consideration of the meaning that such "facts" have for the people who are in the context being studied.[38] Similarly, a practice that seeks to secure and maintain the intersubjective experience of good governance without understanding the meaning of its actions for those affected is likely to misapprehend the proper way to proceed.

The importance of hermeneutics to public administration, then, derives from its focus on intersubjectivity and from the application of its interpretive methodology to the study of human activity and its products ("texts" of all sorts). That is, because hermeneutic theory rejects both the sufficiency of "objective" study and the relativism of those who focus only on the subjectivity of social actors, its primary methodology (the "hermeneutic circle") seeks a "fusion of horizons"[39] between the social/historical context of the scholar and practitioner and the broad and local intersubjective meanings of administrative structure, behavior, process, and procedure for those affected by administrative action. Hence, hermeneutics allows scholars and practitioners both to "read" the impact of their theory and practice on the intersubjective experience of good governance and to understand how those encountering administrative theory and practice "read" public administration.

More specifically, given public administration's purposes and subject matter, the usefulness of hermeneutics to theory and practice in public administration derives from the fact that hermeneutic theory rejects not only the behaviorist assumption that "covering laws" are discoverable and the institutionalist preconception that certain dynamics are everywhere at work at all times, but the view that everything is subjective as well. It recognizes, of course, "a characteristic common to all interpretation: no proposed reading [theory or preconception] can establish itself beyond all possibility of challenge."[40] But rather than suggesting that this truism leads to relativism or nihilism, hermeneutics points to the reality of the socially constructed, intersubjective meaning of institutions and their practices and seeks to understand that meaning through the practical exegesis of the "hermeneutic circle," a methodology involving

> the subtle interplay of global and local interpretation. One the one hand, it is "the very first principle of construction to read the whole instrument before pronouncing on the interpretation of any section, and still more of any single word in that section." Global interpretation takes precedence. On the other hand, this global interpretation must itself be a function of the readings of the parts. This double movement of generation of the global by the local and revision of the local in the light of the global forms the core of the hermeneutical dynamic.[41]

By employing this sort of methodology, theory in public administration can avoid both universalizing (globalizing) the meanings it gives to administrative structure and practice, and marginalizing the meanings that others give them. Hence, it allows scholars and practitioners to understand the intersubjective meanings (actually there are no other kind) that are already embedded not only in the normative, institutional, and behavioral traditions of administrative agencies and the society at large, but in any impacted "local intersubjectivity" as well, in order to determine what these meanings indicate with regard to suitable administrative structures, practices, and procedures.[42]

Equally useful to administrative theory and practice is the evolutionary component of hermeneutic theory. Hermeneutics rejects behaviorist and institutionalist attempts to ground normative theory in an "essential" human nature, or a universal "instrumental value theory," or the cultural or institutional "embeddedness" of human action, or the general categories of reason.

Instead, it begins with the notion that normative theory develops (should develop) from intersubjective needs, interests, and desires and that empirical theory issues from particular intersubjective understandings and traditions. Thus hermeneutics stresses that a particular society's historical experiences imbue its symbols, practices, behaviors, institutions, norms, and roles with particular and changing meanings. These meanings change as they "grow up" with the conceptual systems, institutional arrangements, practices, and norms, developed by that society to cope with its particular historical experiences.[43] Consequently, "even prior to any interpretation whatever, [our] perception[s] . . . [are] not a structureless chaos,"[44] and these perceptions (and the normative and empirical theories they inform) may lag current experience and hence current meanings, especially the meanings emerging within local (e.g., marginalized or "expert") intersubjectivities. This insight is important to administrative theory and practice as it points to the necessity of critiquing its normative and empirical theories and methodologies to insure that they comport with meanings currently held both broadly and locally, and with critiquing those currently held meanings to ensure that they are ontologically rather than ideologically grounded.

As useful as hermeneutics insights and methodology are, however, they are not sufficient to the purpose and subject matter of public administration. First, it does not provide the sought-for synergy among methodologies. Rather it eschews traditional social science by denying the usefulness of its insights and the efficacy of its methods, and hermeneutics by "[implying] an observable arrangement of people's affairs that contrasts with characterizations of people's activities deriving from assumptions, intuitions, or introspection."[45] Second, like institutionalism and traditional social science, it is ultimately engaged in elaborating a particular culture rather than opening avenues to truly novel approaches. Because it focuses upon language and symbol use (and nonuse or silence) in discourse as contextualized in the sociocultural traditions surrounding its use,[46] and upon the meanings embedded in a society's complex of normative, institutional, and behavioral traditions, hermeneutics does not engage in the sort of critique that might suggest fundamental restructurings of meanings or radically different ways of reconceptualizing current terms should their current usages fail to secure or maintain the intersubjective experience of good governance. Hence, it too contents itself with the mundane assemblage of a "mixture" of partial and necessarily constrained improvements.

Some attempts have been made to rehabilitate this aspect of hermeneutics by "foregrounding and appropriating one's own fore-meanings and prejudices . . . [and thus becoming aware] of one's own bias, so that the text can present itself in all its otherness and thus assert its own truth against one's own fore-meanings."[47] This "ontological hermeneutics" seeks to free the "reader" of prejudice by examining historically inherited and unreflectively held prejudices and altering those that disable our efforts to understand others. Thus, it is "based on a polarity of familiarity and strangeness . . . in regard to what has been said: the language in which the text addresses us, the story it tells us. . . The true locus of hermeneutics is this in-between."[48] This relatively new rendering of what hermeneutics is all about, of course, positions its endeavor close to the locus of administrative theory and practice.

However, as Goedel demonstrated, turning "systems of thought" back upon themselves is problematic, because all such systems are themselves subject to historical conditioning and ideological deformation.[49] Consequently, any critique of a system of thought by that system of thought faces the dilemma that the preunderstandings undergirding the critique can be distorted by the specific cultural horizons of that system. For the same reason, confronting a strange system of thought may alert us to the need for a critique of our own, but interpreting that strange system for purposes of critiquing our own system seems to involve necessarily the conceptualization (appropriation) of its terms according to our own preconceptions. True critique, then, seems to require some third system of thought to bring the strange and familiar into true confrontation.

A third reason that hermeneutics is useful but insufficient to administrative theory and practice involves the extent to which it fails to take into account the effects of a society's power structures upon the historically and culturally transmitted meanings it reads. The hermeneutical assumption, for example, that the endurance of an institution (a "text") demonstrates the ontological groundedness of the meaning it expresses overlooks the degree to which the institution's perseverance might be due to power and ideology. However, given the nature of public administration's subject matter and the locus of its endeavor (however close it might be to those of the new hermeneutics), it must in theory and practice recognize that the meanings expressed by cultural texts (e.g., institutions) are not necessarily expressions of ontologically grounded interests and needs, but may rather express both an ideology (i.e., an abstractly grounded, or utopian, or teleologically grounded set of interests and needs) and the power relationships that ideology entails. Any interpretation that does not take such ideologically generated meanings into account runs the risk of uncovering only a society's false consciousness.

Finally, the hermeneutical assumption that there is a historical essence to meaning must give administrative theory and practice some pause. In delineating the disciplinary matrix of public administration, we pointed out that concepts develop "as in spinning a thread we twist fibre on fibre. And the strength of the thread does not reside in the fact that some one fibre runs through its whole length, but in the overlapping of many fibres."⁵⁰ Hence the notion that the meanings carried by a common word must have a common essence is not supported by ordinary language. So a common word need not denote an essential meaning but only a grouping or twining of resemblances among the uses of the term.

Now, the particular way that any given concept becomes twined together is determined in concrete social practice by both the ways a word is used and the ways that others respond. That is, it is "only in the stream of thought and life [that] words have meaning,"⁵¹ and using words in particular ways is thus part of a particular "form of life." Put another way, the meaning of a word "explains the use of the word" in the language games embedded in the concrete customs and practices of a "form of life,"⁵² and as customs and practices "drift," the uses of words are not "fixed" in the strict sense of the word. Novel uses may be presented and "negotiated into" the language game, but to be successfully "negotiated in" (and hence to constitute a "legitimate" use of the term), all such usages must be intelligible, and intelligibility rests upon "family resemblances" to the ways the term has been previously used. Thus, although there is not any single "essence" legitimating each specific use of a word, there is nevertheless a "genealogical" relationship linking each intelligible use with all other current and previous uses. In brief, all uses of a term (in order to be intelligible) must be linked through connections that may fade, but that still provide a distinguishable trail.⁵³ However, this linking of uses does not suggest that there is a "historical essence" to the meaning of the term being used; rather, it describes an evolution of meaning thorough time and space. During the process of this evolution, differences of degree may become differences in kind; that is, the "essence" of a term (its meaning in a particular place at a particular time) may change over time and be different across contexts.

The Role of Critical Theory

Thus, while all three approaches are useful in understanding how to retain the attachment among theory, practice, and the reality of an intersubjective experience of good governance, none alone is sufficient. A synergy among these diverse and sometimes opposing approaches is therefore desirable, given the nature of public administration's subject matter, its purposes, and its locus of scholarly and practical endeavor. What's missing, in other words, is a method or agent for making these approaches and their methodologies work in concert. That is, as traditional social science, institutionalism, and hermeneutics all reveal information that is important to administrative theory and practice, and as all reveal important dimensions and insights into social behavior and intersubjective experience, their application at that locus in concrete situations where administrative theory and practice go to work should involve a complex, integrative, and reflexive agent that employs their insights and methodologies in the mix and manner that is most contextually appropriate to securing and maintaining the intersubjective experience of good governance. This method or agent of broad and reflective synergy must be capable of synergy with the plurality of traditions within traditional social science, institutionalism, and hermeneutics; the diversity of ideals and paradigms within those traditions; the constructed and contingent character of the traditions; and the communities of discourse within those traditions. Hence, maintaining the connection among administrative theory, institutional practice, and reality requires a special kind of method, a method of critique that is connected to practice and engaged not in the "elaboration and affirmation of a culture," but in the identification and affirmation of practical human needs and interests regardless of how at variance with existing "ideologies," behaviors, meanings, and current institutional practices its findings and thinking might be.[54]

It is our contention that the necessary sort of critique may be provided by a critical theory that is informed by an evolutionary rather than a radically democratic paradigm of the sort that has guided critical theory thinking from its beginnings in the Frankfurt school. In thinking about why this is so, it is helpful to consider the kinds of activity that count as theorizing in social science. Briefly, we might distinguish among activities that orient us toward political, social, and economic phenomena in various ways; activities that present and evaluate particular theoretical statements that are based upon those orientations (metatheory); and activities that identify errors in the formulation of a theory, revise the theory to correct the errors, and then reformulate and test the results (research programs).[55] Orienting activity articulates the conceptual, ontological, and epistemological foundations to be employed in the description, analysis, and explanation of social phenomena, thereby defining what theoretical problems are possible and how they might be investigated.[56] Metatheory relates a set of concepts (usually suggested by an orienting strategy) and a set of assertions relating those concepts to each other in the form of "propositions," "axioms," or "causal models," in order to describe or explain some social phenomenon. Theoretical research programs involve the articulation and evaluation of interrelated theories, each often having developed through adjustments made to some original theoretical statement on the basis of testing.[57]

As both bane and complement to these activities, critical theory is first and foremost an activity calculated to insure that we remain persistently self-conscious of the fact that neither the social phenomena, nor the institutional structures, nor the meanings that we study, nor the concepts, epistemologies, or methods that we study them with, nor the resulting models, or explanations that we derive in the course of our study, are ontological givens. This critical self-consciousness is important to maintain because it is just when these values, presuppositions, meanings, and habits of thought and behavior are taken as ontological givens that theory and practice divorce from reality. And once so divorced, they are at risk of becoming the source of political, social and economic subjugation, marginalization, and alienation. They become the source of the divergence between what a given set of institutions promises and what it delivers; and if it delivers on the whole, they are the source of the remaining discontent, the continued subtle sense of meaninglessness and the ultimate resignation to something being not quite right that expresses itself in the idea that the remaining subjective discontents of individuals are unrealistic. Critical theory, then, sets itself the task of providing a means of identifying and critiquing these fundamental ideologies, conceptualizations, epistemologies, meanings, and practices in order to focus attention upon them, to explore the interests and needs that gave them dominance, to provoke their reconsideration in light of current interests and needs, and to encourage thinking about how to change in order to reunite theory, practice, and the intersubjective experience of good governance. Thus critical theory "is not just a research hypothesis which shows its value in the ongoing business of men; it is an essential element in the historical effort to create a world which satisfies the needs and powers of men."[58]

Critical Theory as Synergistic Agent

In explicating "critical theory" as a synergistic agent, we are aware that the term covers a robust, diverse intellectual tradition that cannot be characterized by a closed set of elements or reduced to a single set of shared characteristics.[59] Nevertheless, it is undeniable that the term is used coherently. That coherency, of course, is the result of certain "family resemblances" among the different uses of the term,[60] and it is to those resemblances that we must point in order to meaningfully employ the term for our purposes here. Consequently, what follows derives from an understanding of critical theory garnered from the writings of critical theorists as read in concert rather than as in contradistinction to one another. By choosing not to reject one school of critical theory in favor of another, we believe we are being true to the tradition itself, because it abjures any reduction to essences and denies the efficacy of any single right way of doing things in every context. At the same time, our understanding of the term is not intended to represent the sort of interpretational endeavor that seeks to merge diverse ideas and themes under a single rubric, suggesting that the heretofore robust intellectual and interdisciplinary interaction that is the hallmark of critical theory might ideally result in some kind of consensus or strong, concise statement about its content and substance. It is only meant to represent the fact that the usages of the term "critical theory" exhibit an integrity and coherence that renders the tradition as a whole advantageous to theory, practice, and research in public administration. With these caveats in mind, it is nevertheless clear that the importance of critical theory as a synergistic agent in the theory and practice of public administration is unmistakable, given public administration's locus and the nature of its subject matter.

The Synergy with Traditional Social Science

Hegel articulated the idea that both thought and social reality (theory and practice) progress through a series of contradictory (and hence unsatisfactory) stages to a final absolute wherein all contradictions are resolved. Critical theory denies the possibility of the final stage. Consequently, it seeks to provide a dialectical method of discovering and rediscovering "better ways" to develop people and transform society in always fluid contexts. These "better ways," of course, must include dialectical methods for continuing self-critique so that such a process may proceed indefinitely as circumstances change.[61]

Such better ways cannot be discovered by the different forms of empiricism alone (whether from a behavioral or institutional perspective), as their goal is not social transformation but the better working of particular elements within the existing (and taken as given) social structure, and often reveals only appearance, false consciousness, and ideology.[62] Nevertheless, critical theory currently accepts what empiricism does reveal as part of the totality (the overall historical, social, cultural, and natural reality) within which the "better ways" must be found. Moreover, even during Hegel's time, it was considered by many to be self-evident that a theory of society could engage in critique only insofar as it was able to discover an element of its own critical viewpoint within social reality.[63] To discover the better ways within this totality, then, requires that we understand the character of the social totality, the unmet ("contradicted") needs and interests within the totality at any given time, and the most efficacious ways of exerting conscious control over both in order to proceed to a meaningful (ontologically based) transformation. Thus a synergy among reflection or dialectical reasoning, empirical description, and instrumental reasoning is especially valuable. What must be avoided, of course, is positivism's separation of *facts* and *values*, as this separation precludes the formulation of sound, accurate descriptions and explanations (i.e., theories of social, political, and economic processes that are firmly rooted in reality).[64]

Negative dialectics is a time-honored critical approach that can prove effective in accomplishing this synergy. Negative dialectics involves the testing of every social construction (e.g., every theory, "fact," forefronted object, and discerned pattern of behavior "identified" and studied by traditional social science) by negative scrutiny in order to reveal whether these constructions are ontologically grounded (i.e., grounded in the broad and local contextual needs and interests of people) or the product of mere ideology (*doxa*). Negative scrutiny is accomplished by initiating immanent encounters of the particular with a society's dominant, apparently neutral discourses, ontologies, axiologies, and epistemologies. These immanent encounters (through, e.g., fantasy in science, surrealism in art, "reflexive" dialectics in philosophy and discourse, advocacy for the marginalized in politics) focus upon deconstructing the individual identities, the dualisms, the meanings, and the sociopolitical organizations that have been constructed through both empiricism and the dominant culture.[65] Such deconstruction is expected to throw into sharp relief the antinomies between a society's theory and practice (promise and performance), thereby engendering a consciousness of the dynamics establishing and maintaining the antimonies, encouraging the bringing of these dynamics under conscious control, and spurring societal transformation. In this way, negative dialectics intends to accomplish a maximal adaptive responsiveness of both people and their constructions to all contradiction, paradox, and idiosyncrasy.[66]

But the negative dialectic is intended to function as more than a mechanism for testing all social constructions. It is intended to play an integral role in formulating those constructions as well. That is, it is meant to constitute a social practice that is revelatory of contextual truth and therefore suggestive of how social constructions might be transformed or reconstructed to take different contexts (including different subject matters) into account. Traditional social science, of course, is one such construction. As such, it is open to transformation and reconstruction, and negative dialectics, as revelatory of contextual (particular) truth, is capable of playing a significant role in such reconstruction.

Consider, for example, the scientific method. Broadly, it consists of definitions, hypotheses (illustrations), theory (beliefs), and proof (experiments). Each of these requires social (not just professional) construction. For example, what counts as a fact, what counts as an anomaly, and what counts as a credible experiment all depend upon both the broad as well as the local social processes within which professional scientists operate. Social science, therefore, "must be understood as constituted, not just shaped or influenced or interfered with, by its social, political context."[67] That is, it must be understood as the construction of particular groups acting in concert with the politics, culture, socialization, history, media, literature, and religion of a given society.[68] By drawing attention to the political, economic, and social interests that influence the construction of definitions, theories (beliefs), and successful scientific work,[69] negative dialectics plays an important role in constructing what counts as both *good science* and *bad or failed science* and therefore in not only negative but positive critique.

These examples illustrate the reciprocal or synergistic "criticism of the universal and of the particular; identifying acts which judge whether the concept does justice to what it covers, and whether the particular fulfils its concept."[70] The universality of the concept science cannot exhaust (cover completely) what we encounter in the particular cases of those doing science now or in the future, given the fluidity of contexts and the ontological diversity of subject matters. At the same time, the concept contains the idea of a socially constructed ideal process and procedure that is impossible to achieve in the particular (e.g., the control of all external variables) that informs the particular. It is in this sense that critical theorists say "that which is cannot be true,"[71] and that traditional theory is wrongly fixated on the idea that theory is some set conglomeration of propositions and facts about an object ordered in a particular form such that all knowledge may be deduced from a few basic postulates.[72] Negative dialectics is intended to overcome these misapprehensions concerning the "facticity of the given"[73] by acting as both a critique revealing "the deficiencies of conceptual thought [and any other social construct] . . . [and as a] critique [revealing] the inadequacy of objects [e.g., particular ways of "doing science"] with respect to concepts [the broad construction of "doing science"]."[74]

The Synergy with Institutionalism

Critical theory traditionally focuses on the role of institutional factors in holistic social dynamics and on what a study of those institutionally affected dynamics reveals about the totality (the overall historical, social, cultural, and natural reality) within which the better ways must be found. Thus it has always been interested in what institutionalism describes, though not what institutionalism prescribes or offers as explanation except insofar as these reveal the relationships of power within a society. One of critical theory's primary concerns, for example, is the extent to which the ideology and characteristics of administrative institutions restrict the exercise of free choice to smaller and smaller, more and more marginal, spheres of public and private life, directing our efforts through narrowly technorational definitions of what our problems really are, triggering routinized, legalized, and rationalized *appropriate responses*, and either imposing sanctions or denying aid and reward should we fail to regulate our behavior accordingly. Thus the conclusion of many critical theorists is that "the pattern of all administration tends of its own accord . . . toward Fascism."[75]

For the reasons delineated above, however, the better ways sought by administrative scholars and practitioners cannot be discovered through institutionalism alone. Nevertheless, *immanent critique* and *praxis*, as developed by critical theorists, can prove effective in accomplishing a synergy between institutional approaches and critical theory that renders institutionalism more helpful to administrative theory and practice.

Immanent critique, for example, "confronts the existent, in its historical context, with the claim of its conceptual principles, in order to criticize the relation between the two and thus transcend them."[76] Thus, immanent critique is in an important sense Janus-faced. Its one face confronts the disparity between values and reality. Its other gaze pierces the social veil (including the veil of values) to expose what is socially given as not ontologically constituted but constructed for certain purposes (i.e., to serve certain interests and needs), and to criticize what is given by showing what it has the potential to be. On the one hand, immanent critique examines the conceptual principles and values of an object (i.e., a social construction) "and unfolds its implications and consequences. Then it re-examines and reassesses the object . . . in light of these implications and consequences."[77] In this sense, critique proceeds *from within*, by pointing out where extant social processes and outcomes depart from the foundational norms and ideals of a given social order. On the other hand, it critiques the foundational norms and values of a society in order to reveal the extent to which these very foundations render impossible the closing of many ideology/ practices gaps. Thus it is a method of critique that works within existing institutional forms and within the existing categories of thought, revealing both their problems and their possibilities. It is both a negative and a positive critique, intended as a tool for identifying and keeping the scholar and practitioner focused upon the tensions between the socially given and its possibilities.

Thus, as applied to our argument here, immanent critique proceeds along both a normative and an empirical or pragmatic dimension. Its normative dimension focuses on explicating institutional values and pointing to any failures by specific institutions to live up to their own stated ideals. Its empirical dimension, on the other hand, makes explicit an institution's ideology (i.e., its ontological and normative beliefs, its tacit assumptions, its habits of thought, its practices and power relationships) and subjects it to pragmatic contextual criticism by pointing out the gaps between what obtains in practice and what the theories (ideologies) an institution employs to justify its activities claim will be the results. In this way, both institutional practice and institutional values are critiqued by way of their practical implications. However, the point of the critique is not simply negative. Immanent critique is intended to function as more than a mechanism for testing all institutional practices, values, attitudes, and beliefs. It is intended to play an integral role in formulating those practices, values, and beliefs as well. That is, like negative dialectics, it is meant to constitute a social practice that reveals contextual truth and therefore suggests how institutional practices, attitudes, values, and beliefs (ideologies) might be transformed or reconstructed to take into account whatever particular human interests and needs emerge from different contexts.

It is important to remember, along this line, that critical theory does not necessarily disdain ideology per se, "but only its pretension to correspond to reality"[78] and its tendency to preclude or marginalize any consideration of what might otherwise be. In fact, under the proper circumstances, ideology may in fact express a truth—just not necessarily the truth it thinks it is conveying. Its truth-content may lie in some indirect suggestion of what is wrong now and what could constitute an improved situation. Thus, to critical theory, "the fatal part of positivist ideology is not that it is "untrue." In fact, it is true enough in certain contexts to serve effectively the shared technical interests that we all have in coping with the environment.[79] It is only when positivism steps out of its proper contexts and suggests that "what ought to be is what is,"[80] or that people and their lifeworlds may be treated as instrumental things rather than as ends in themselves,[81] that critical theory holds it up for disdain. Hence, ideology (as in the case of religion) may function as "not mere domination; like the traditional authority embedded in historical values and hierarchically structured power relations, but the product of communal negotiation of needs in response to the threats of nature."[82] In other words, the problem is a dominating ideology that both distorts the conditions of social life and justifies the status quo against intersubjective interests and needs.[83]

Immanent critique is directed at the preclusion of ideological domination by suggesting not only when an ideology is destructive, but also under what conditions it may be useful and in what ways it may suggest what is possible. Thus, the role of immanent critique involves a search for the best application (if any) of an ideology as recognized by the intersubjective experience of a community. It is in this sense synergistic with the ideologies (values, attitudes, and beliefs) held by institutions and expressed in their structures, practices, and procedures. However, in providing the critique necessary to these ends, and in order to critique institutional character by way of its practical implications, immanent critique cannot remain simply an abstract endeavor; it must be translated into practice. It is for this reason that praxis plays an important role in the synergy between institutionalism and critical theory.

Praxis embodies the Marxist belief in "the unity of thought and action, theory and practice";[84] it is an activity aimed at transformation and self-development.[85] Rather than thinking of practice as "doing" and theory as "directing" through a set of interrelated propositions about both what our goals should be (indeed, what goals are possible) and how we should go about attaining them, praxis seeks to discern proper ends and proper actions by fusing action and deliberation in context.[86] It takes advantage of the fact that as we actually go about achieving ends, we tend to develop theory in those particular contexts where routine (thoughtless) practice either produces novel problems or encounters obstacles that it cannot overcome, and that novel practices arise in those contexts where theory fails to produce desired results. In these situations, the most productive approach most often involves both an alteration of what we do as we think about what we want to achieve, and an alteration of exactly what we want to achieve as we think about how we might go about doing things effectively. Praxis, therefore, involves an integration of thought and action, theory and practice, so as to achieve in "one unified process" a continual deepening and broadening of our contextual understanding so that we might achieve our ends and enhance our theoretical understanding by continually reconceptualizing exactly what we must do.[87]

As applied to administrative theory and practice, praxis and immanent critique together can provide the desired synergy between institutionalism and critical theory. Praxis provides an ongoing confrontation with context and an impetus toward reconceptualizing ends and means, and immanent critique provides a means of achieving reconceptualization by analyzing the extent to which institutional ideology and the values, practice, theory, and results cohere, and the extent to which institutional ideology of institutional values, attitudes, and beliefs where that is all the more effective when immanent critique is employed to overcome institutional preconceptions, to test institutional ideology (practices, values, attitudes, and beliefs), and to play an integral role in reformulating those practices, values, and beliefs as well. Employing the two together, in other words, establishes an effective synergy not only between institutionalism and critical theory, but between institutional theory and institutional practice as well.

The Synergy with Hermeneutics

As explicated above, hermeneutics requires some third system of thought to overcome its tendency to elaborate and reinforce a particular culture even when it is confronted with a strange system of thought. This tendency results from the fact that hermeneutics is a social construction and therefore, like all social texts, both produces and articulates a society's broader ideological interests, social structures, and history. Its interpretations of texts (even very strange texts that indicate the need to reconceptualize before they can be understood) are therefore conditioned by and inscribed within a sociocultural tradition (one it has itself played a part in constructing). Critical theory can prove helpful in solving this problem because its system of thought contains both a critical form of discourse analysis that involves a deconstructive reading and interpretation of a text, and a constructively oriented *discourse ethics* calculated to aid in the understanding of strange systems of thought. Hence, it may provide a synergy with hermeneutics that reveals the sociocultural horizons (limits) of textual interpretations and suggests how to go about working out more clearly what people with strange systems of thought are thinking.

More specifically, hermeneutics (like all social constructions) is a particular text, invented to accomplish certain ends and to have particular ideological and practical effects. Hermeneutics (again, like all social constructions) is, of course, dynamic and continually subject to innovation and reconstruction. Nevertheless, to be coherent to those within the sociocultural milieu that constructed it, hermeneutics must remain connected (e.g., through family resemblances in how it uses its terms) with conventional discourses and the preconceptions those discourses reflect. This requisite conditions necessarily the ability of hermeneutics to understand strange texts. Consequently, it continuously runs the risk of reinterpreting or appropriating strange texts in ways that render them commensurate with the sociocultural traditions (e.g., the conceptual and categorical systems, epistemologies, and values) of a given society. Critical discourse analysis provides a method for getting over this problem. By employing interdisciplinary techniques of deconstruction, it can uncover how and to what extent any hermeneutic endeavor is reconstructing in its own terms (i.e., is appropriating) the terms of strange texts and so misunderstanding the values, social identities, and social relationships that those terms embody. In the process, it can reveal how any particular hermeneutic endeavor is shaping what counts as knowledge and sound practice in hermeneutics theory (discourse) and practice (i.e., in the hermeneutic text), thereby suggesting what sort of changes (if any) are desirable in either or both. In its negative (or deconstructive) mode, then, critical discourse analysis can help hermeneutics distance itself from its own discourse and text (theory and practice) by disrupting and rendering problematic any culturally conditioned themes and preconceptions that hermeneutics may bring to its interpretive endeavors; and in its reconstructive mode it can point out how hermeneutics can be used not only to assert and maintain a given knowledge but also to resist and critique that knowledge in ways conducive to a "better reading" of strange texts.

But, having been deconstructed, and having distanced itself from its own text and pointed toward helpful change in its theory and practice, how is hermeneutics to finally arrive at a better reading of strange texts? As is the case with all social constructions, the best interpretation of a text is invented by the intersubjective experience of a community, and in the case of strange texts, that community must include those employing or expressing the strange text in their sociocultural practices. That is, an interpretation of a strange text must be not just coherent but comprehensive as well. It must "make sense" as a "compelling account" to all participants within the community of discourse, and that community must include those whose text is being interpreted.[88] Critical discourse analysis, then, sets up such an inventive process by allowing us to make explicit the socioculturally constructed assumptions and motivations behind both hermeneutics and any particular hermeneutic endeavor, thereby moving us toward a better "reading" of strange discourses as judged by intersubjective experience (i.e., the best interpretation as recognized by the broadest community of experience, including the "community of experts" on whatever sort of discourse is at issue—e.g., administrative theorists and practitioners—and the community of those expressing the strange discourse).[89] For the actual "best reading" to occur, however, we need a method for introducing "strange" intersubjective experiences (and their strange texts) into an interpreting community's "normal" intersubjective experience and discourse, and this method should require that those strange experiences be given serious consideration and effectively filter out discourses that seek to dominate or marginalize those experiences.

"Transcendental hermeneutics" or "transcendental pragmatics"[90] is the critical method that can accomplish these ends. Transcendental hermeneutics provides a set of ethical guidelines for discourse that are calculated to reveal the "structures of mutual understanding that are found in the intuitive knowledge of competent members of modern societies."[91] These guidelines include

> the [selection of] a comprehensible expression in order that the speaker and hearer can understand one another; the [intention] . . . of communicating a true propositional content in order that the hearer can share the knowledge of the speaker; the [sincere desire] . . . to express . . . intentions truthfully in order that the hearer can believe in the speaker's utterance (can trust him); finally, the [selection of] . . . an utterance that is right in the light of existing norms and values in order that the hearer can accept the

utterance, so that both speaker and hearer can agree with one another in the utterance concerning a recognized normative background.[92]

Following these guidelines engages the speaker and the listener (the strange discourse or text and the interpreter) in an "ideal speech situation" (an inclusive "public space" of interpersonal communication and interpersonal respect) that promotes a "rational" (nonimposed) convergence of behaviors and practices (including textual interpretation).[93]

Of course, the realization of such an ideal reading depends upon the participants' normative predisposition to sincere communication and their good faith in carrying out those predispositions. Often administrative bureaucracies, for example, read and discourse with their audiences in ways that engender false or imposed agreements about what is meant and expressed. But this does not deny the possibility of a best interpretation; the ideal speech situation is what should be done to arrive at a best interpretation, not an ineluctable phenomenon. It is a way of arriving at "true" rather than false consciousness of a strange text on intersubjectively transmissible (i.e., nonsubjectivistic and nonobjectivistic or imposed) grounds.[94]

In brief, a better reading (i.e., a consensus reading) of a text can be erected upon a shared language whose development is facilitated by discourse analysis and discourse ethics. However, such a shared language requires more than discourse analysis and discourse ethics.

Suppose, for example, that after we engage in both with those exhibiting a strange text, there emerges no consensus on a better reading can be established; we have tried to devise a language among us that embodies a mutual understanding of the text, but we have failed. We then find ourselves in a situation "where two principles really do meet which cannot be reconciled with one another, [and] each man declares the other a fool and an heretic."[95] At this point, "at the end of reasons, comes persuasion";[96] and the success of persuasion depends ultimately upon a shared "form of life."

"To imagine a language means to imagine a form of life."[97] That is, language is necessarily embedded within nonlinguistic behavior that is itself grounded in biological proclivities and capacities as conditioned and transformed by historically specific cultures. These intersubjective experiences of constructed social reality constitute the frameworks within which human dynamics occur and are interwoven with inventing what human beings will be (how human dynamics will express themselves) in any given context. Thus, the human form of life is fundamentally cultural and biological in nature. It emerges slowly and uniquely; biological proclivities and capacities are molded, suppressed, transformed, and further structured by culture, experience, and the acquisition of new language-games. Consequently, human forms of life have both an ontological and a derivative constitution; and many human forms of life come, exist simultaneously, and go through history. But they are all human forms of life, though they are diverse and have no one defining core; and it is ultimately to that which is human in the forms of life of both the interpreter and those exhibiting a strange text, that persuasion must be addressed.

Once again, critical theory is helpful here. Human nature is both ontologically constituted and derivative of social ideology, institutions, and values so far as critical theorists are concerned.[98] The ontological constitution of human beings includes a primal fear of nature,[99] the "need to relate to others . . . [the need] to feel rooted in a world they consider their own . . . [the need] to transcend their feelings, [the need] to be a creature either by creating or by destroying . . . [the need] to have their own sense of identity . . . and [the need] to have a frame of orientation that gives some meaning to the world they live in."[100] It also includes a tendency to imitate and identify with those most like themselves,[101] a consequent fear of nonconformity,[102] a "biological necessity" to both organize and elaborate the complexity of that organization, and an instinct for negativity, refusal, and rebellion.[103] Moreover, human nature is inclined toward both self-sacrifice, domination,[104] and the "distinctly human pleasures" of ecstatic self-forgetting and a certain longing for transcendence.[105] Most recently, critical theorists have included within the ontological constitution of human nature not only certain natural technological interests regarding nature, but certain social and communicative interests regarding social reality and an interest in being emancipated from the constraints of both nature and social reality that work against people's interests as intersubjectively understood.[106]

But even as they recognize these ontological proclivities, critical theorists simultaneously avoid offering any definitive concept of a human nature that might be employed as a generalized norm transcending the experiences of particular peoples in particular contexts. They avoid this idea of a human essence because they recognize that while human nature has a dynamism of its own, lent to it by the proclivities and capacities of human nature and constituting an active factor in the development of the individual and the evolution of the social process, there is no biologically fixed human nature.[107] Rather, the socially expressed human nature that we encounter in great diversity is a third-order construction, constructed from a social reality that is constructed from our constructions of nature (including our own second-order constructions of our own proclivities and capacities). Because both nature and social reality are in this sense constructed intersubjectively, it is precisely those intersubjective experiences and processes of construction that constitute the subject matter of critical theory. And it is those constructions that mould and transform human forms of life to produce domination and thereby unnecessarily suppress or distort human proclivities and capacities that critical theorists challenge through negative dialectics, immanent critique, praxis, and transcendental pragmatics. And it is to those proclivities and capacities that critical theorists point ultimately in order to establish a consensus reading of the texts around us for purposes of both critique and improvement.

The Role of Critical Theory and Public Administration

So far we have argued that the methods of traditional social science, institutionalism, and hermeneutics are all useful but individually insufficient to theory and practice in public administration. And we have argued that negative dialectics, immanent critique, praxis, and transcendental hermeneutics can render each more useful to administrative theory and practice by keeping scholars and practitioners conscious that objects, patterns, and relationships can always be forefronted and analyzed through multiple conceptual, normative, theoretical, and historical perspectives, and by encouraging multiple constructions and analyses through an ongoing critique employing the methods of negative dialectics, immanent critique, praxis, and transcendental pragmatics. To complete the argument, we must demonstrate that it is appropriate to apply the methods of critical theory to the subject matter of public administration. We must be sure, in other words, that we are not engaged in the sort of mistake that people make when they apply the methods of natural science to questions of morals or questions of fundamental conceptualization (e.g., is a fetus a human being? should it be treated as one?). As useful as one might find employing scientific methods in such contexts for various reasons, they are in the one case inappropriate to the subject matter and in the other inappropriate to the task at hand. Thus we must insure that critical theory does not do violence to the nature of public administration's subject matter and that its use is appropriate to the goals and locus of public administration's endeavor.

As argued above, given the nature and ontological status of intersubjective experiences of good governance, and given the purpose of administrative practice, public administration's central concern must be the transition points or loci of synergy between the individual and the social, between purpose and phenomena, between meaning and behavior, between culture and psychology, between the particular and the general, and among institutions, individuals, and groups. In addition, administrative theory and practice must remain thoroughly embedded in an institutional perspective, seeking to explain both individual and aggregate behavior as the result of individual experience, institutional rules, roles, norms, and expectations; human agency (nondetermination) by experience or institutional character; and the conditioning of individual, group and institutional identities, values, and habits of thought by history, culture, context, and roles vis-à-vis institutions and their needs. It is our contention that critical theory is, on the whole, appropriate to this subject matter and to these tasks. To the extent that it is inappropriate, it can be cured by employing it within an evolutionary as opposed to a democratic paradigm.

First, critical theory's subject matter is the intersubjective experience of nature and social reality, and this renders it particularly appropriate to theory and practice in public administration. Moreover, its ontology is commensurate to this subject matter as understood by public administration. Among its foundational insights, for example, is the understanding that humans share "categorical distinctions among the objective, social, and subjective worlds"[108] and that social reality is constructed intersubjectively in a different way than is ("objective") nature.[109] That is, critical theorists consider nature to be an ontologically constituted, primal, and intersubjectively shared subject matter that is largely not factual except as humanly constructed, and to the extent that it is internally structured or determined prior to the knower's engagement, it is nevertheless sufficiently complex to be grasped intersubjectively in many different ways. Hence, critical theory is appropriate to the subject matter of public administration.

Next, critical theory takes seriously the idea that all concepts and categories, all systems of thought (critical, analytical, or synthetic), all methodologies (e.g., the scientific method), and all institutions are socially and historically constructed, though grounded in certain human capacities, proclivities, and interests concerning "genuine communication" for technological, social, and emancipatory ends.[110] Meaning in particular is constructed; it is not there as part of an intersubjectively shared subject matter. Rather, it is a matter of social negotiation accomplished largely through language. Consequently, critical theorists distrust all totalizing notions as, for example, the idea that there is an essential human nature that might be employed as a generalized norm transcending particular people at particular times, likely in practice to marginalize, denigrate, or disadvantage. Thus, oppression and social injustice are often the result of social and historical constructs. All such constructs are addressed to historical and not contemporary conditions, and all privilege some consequently. However, as all of them are socially constructed, they all may be socially deconstructed and reconstructed in better ways as worked out through ongoing critique employing negative dialectics, immanent critique, praxis, and transcendental pragmatics. Hence, critical theory is appropriate to the tasks of administrative theory and practice.

Finally, the methodologies of critical theory allow the necessary ongoing adjustment of administrative theory and practice to the changing intersubjective experience of good governance. These methods recognize that the results of all forms of intellectual enquiry are provisional only. There are no fixed or reliable findings and so no fixed or reliable truths (including the statement that this is so, as reliable truths may be for specific contexts at specific times). All methodologies, all theorizing about the findings they produce, and theoretically and methodologically grounded practices are conditioned significantly by the interests, theoretical perspectives and ideological commitments of scholars and practitioners, and therefore require ongoing critique and alteration to keep pace with the flux in intersubjective experience.

The Advantage of an Evolutionary Paradigm

Critical theory has always assumed that the recognition of human diversity and the meaningful recognition of genuine human interests can be accomplished only through a genuine, all-inclusive, radically plural form of democracy. It is thus constituted by diverse attempts at conjoining reason and free speech with the ultimate goal of institutionalizing a form of democratic social organization that empowers each individual to such an extent that freedom, justice, and reason are real possibilities.[111] Critical theorists expect that the meaning of each of these objectives (and each of their constituent terms) will be constructed, deconstructed, and reconstructed socially and historically. Still, the vision of a nondualistic, nonhierarchical polis of individually empowered citizens constructing, deconstructing, and reconstructing society and themselves in a relentless maximal synergism with each other and their environment constitutes a transcendent ideological element or "final cause" in all of critical thinking from start to finish. Thus it always "retained a transcendent or utopian component. . . A commitment to the integrity of the individual, and freedom beyond existing parameters, became perhaps the motivating factor behind the entire enterprise."[112] However, "the equilibrium holding between the immanent and transcendent elements of the project was always tenuous at best."[113]

But if our subject matter is the intersubjective experience of good governance, the question becomes whether there is broad and local intersubjective legitimation, such that all significant political actors, at both the elite and mass levels, believe that a democratic regime is the most right and appropriate for their society, or at least better than any other realistic alternative they can imagine. This legitimacy must be more than a commitment to democracy in the abstract; it must also involve a shared normative and behavioral commitment to the specific rules and practices of the country's constitutional system, regardless of whether it works for a particular people and serves their interests.

Habermas, for example, makes no apology for his confidence that only by employing something close to his ideal speech situation might a radical form of participatory democracy (or at least some synthesis of liberalism and civil republicanism) be obtained.[114] King is equally committed to a "more democratic . . . scholarship where knowledge and understanding are emergent" and everyone is engaged in a "participatory collusion."[115] Her suggestions for "healing" the theory-practice gap are instrumental to that end and intended to empower marginal voices. Finally, the entire point of Zanetti's proposed methodology is to democratize and sociopolitically empower everyone and every community by overcoming the conflicts engendered by principles challenging the moral framework of complete democratic inclusion. In brief, the ingenious suggestions for closing the theory-practice gap offered by Habermas, King, and Zanetti are meant to be instrumental in achieving the same transcendental ideological vision characteristic of critical theory from the Frankfurt school on.

However, as enticing as the ideas proffered by Habermas, King, and Zanetti are, they fall short conceptually because the ideas of a wholly participative democracy, designed to empower marginal voices, the disenchanted, and those otherwise alienated, point toward a single teleological outcome that may or may not be what is best for society. If we reflect on the idea of a wholly participative, minimalist administrative state, we would find rather quickly that such an arrangement could arguably cultivate many of the problems these scholars sought to correct. This problem, which appears both knotty and inevitable in the context of what we understand as critical theory, then forces us to reconsider our positions about society and how critical theory might help us arrive at the intersubjective experience of good governance at the core of public administration. If we instead consider a critical theory unencumbered by such a teleology and able to change over time, we then might argue that public administration's primary role in creating the intersubjective experience of good governance comes from its ability to function both for societal institutions and for individuals. This makes it the primary means to address both the issues and voices of all, including the marginalized, disenchanted, and alienated.

To achieve such an end, we must look to the writings of Thorstein Veblen and recast critical theory from a nonteleological evolutionary framework. This in turn would allow us to reconcile and address both the goals and concerns proffered by the Frankfurt theorists and their intellectual descendants. Particularly, we can then address the ideas of power and emancipation and how the administrative state fits within such a conception of critical theory. In addition, when we reflect on such issues, we then begin to understand where the discipline of public administration fits within the fabric of society and what its legitimate roles are in such a society.

5

Evolutionary Critical Theory

So far, we have demonstrated that public administration both is a discipline and has a clear ontological status. Given the unusual if not unique characteristics of this ontological status, we then found that critical theory has a great deal of promise for creating the necessary synergy to bridge the theory-praxis gaps that currently exist in the discipline. We discovered, however, that critical theory, as it is currently constructed, is lacking, primarily due to its teleology, making any such synergy among institutionalism, hermeneutics, and traditional social science difficult if not impossible. Arguably, the solution to this difficulty would require us to reconstruct critical theory, removing this current teleology and create the potential for an "endogenous evolutionary" approach, while maintaining its methodology and potential.

Fortunately, we need not reconstruct critical theory in a vacuum. The work of Thorstein Veblen affords us a conceptual foundation to then craft a nonteleological, "endogenous evolutionary" approach, reflecting the core of his critical thought. This core of thought, often considered to be a morass of ambiguity and internal contradictions,[1] represents a clear misunderstanding of the critical nature of his work. If we instead consider Veblen's economic theories as critical, a multitude of interesting possibilities emerge. Veblen's core of critical thought offers up differing conceptions of power and emancipation, which arguably lead to a balanced view of administrative institutions and their roles. As a result, we can then reflect on what constitutes a good society and how both administrative institutions and power ostensibly fit within it.

Therefore, by adopting these Veblenian economic ideas, we can craft a remarkably advanced and robust form of critical theory. Veblen's opposition to every teleology and form of determinism, his immanent critique of capitalism, his insistence upon understanding change in evolutionary terms, and his determination that "consciousness" (gained largely through negative dialectics) is the first step in overcoming alienation reveal this body of research as far more than a mere precursor to the Frankfurt school. Most important, adopting these Veblenian ideas can offer a way out of the teleological crisis critical theory continues to experience in trying to close the theory-practice gap and by explaining how and why historical change is not consistently "emancipatory." Incorporating Veblen's theories constitutes a much more dynamic social construction than critical theory as it is currently formulated and applied.

The Nature of Critical Theory

Although critical theorists declare a shared interest in liberating individuals and unleashing human potential, critical theory cannot be characterized by a closed set of elements and defies reduction to a single set of shared characteristics.[2] Consequently, we must comprehend the term "critical theory" much as we comprehend other words such as "game." We must look for the "family resemblances" among the different things that are done as people go about employing the term.

Briefly, these resemblances cohere about providing systematic, immanent critiques of current social conditions that are intended to envision and realize equality and justice in everyday life.[3] The term, therefore, may be grasped best as signifying "ways of proceeding" or "modes of understanding what to do next" regarding oppression and social injustice. Certain that human life is both worth living and can be improved,[4] critical theory seeks to abolish oppression and social injustice, to promote human integrity (self-rule), and individual freedom, and to bridge the theory-practice gap by reconstructing "theory" as "lived experience." In this sense, critical theory is constituted by diverse attempts at conjoining reason and free speech with the ultimate goal of institutionalizing a form of democratic social organization that empowers each individual to such an extent that freedom, justice, and reason are real possibilities.[5] Critical theorists expect that the meaning of each of these objectives (and each of their constituent terms) will be constructed, deconstructed, and reconstructed socially and historically. Still, the vision of a nondualistic, nonhierarchical polis of individually empowered citizens constructing, deconstructing, and reconstructing society and themselves in a relentless maximal synergism with each other and their environment constitutes a transcendent ideological element or "final cause" in all of critical thinking from start to finish.

To these ends, critical theory takes as given both the mutability of human nature and the lack of any determinism in human affairs (material or transcendental). All categories for apprehending anything; all systems of thought (critical, analytical, or synthetic) and discourse; all methodologies (e.g., the scientific method); and all institutions are socially and historically constructed, though recently grounded in certain human interests concerning "genuine communication" for technological, social, and emancipatory ends.[6] Oppression and social injustice are the inevitable results of these social and historical constructs. All such constructs are addressed to historical and not contemporary conditions, and all constructs privilege some consequently. However, as all of them are socially constructed, they all may be socially deconstructed as well.

Generally, critical theorists hope to engender crises of confidence in a society's ideological and institutional framework. To accomplish this, they emphasize immanent encounters of the particular with a society's dominant, apparently neutral discourses, ontologies, axiologies, and epistemologies. These immanent encounters (through, e.g., fantasy in science, surrealism in art, "reflexive" or "negative" dialectics in philosophy and discourse) focus upon deconstructing the individual identities, the dualisms, the meanings, and the sociopolitical organizations "received" from elites through the dominant culture.[7] Such deconstruction is expected to throw into sharp relief the antinomies between a society's theory and practice (promise and performance), thereby engendering a consciousness of the hidden mechanisms of oppression and spurring societal transformation. All critical theory seeks, in practice as well as theory, a maximal adaptive responsiveness of humans and their constructions to all contradiction, paradox, and idiosyncrasy.[8]

Veblen and Critical Theory

Much of Veblen's research parallels many of the central issues that the Frankfurt theorists wrestled with throughout their dialogues. Veblen, much like his Frankfurt counterparts, understood the constructed nature of what was commonly considered "real," including critiques of "economic man" and how empirical analysis of these functioned to affect both causes and effects.[9] This realization, for example, served to reveal how Smith and his contemporaries were using utilitarian philosophy as an instrument of domination. In addition, much of Veblen's work deconstructs a number of economic concepts, including Bentham's theory of value,[10] and shares the view that mainstream, positivist, and instrumental modes of thought are distortions of reality, failing to emphasize needs and purposes over the implicit roles of so-called natural laws.[11] However, we realize that at some level we are reconstructing Veblen through our own perceptual lenses with the intent of reconstructing a critical theory that will serve the purpose of linking theory and praxis in public administration. Furthermore, such a reconstruction serves to achieve the end of casting critical theory as the synergistic agent, achieving the ends we established earlier.

While we are thus aware that as "reconstructive agents" we are stressing and arranging certain elements and dimensions of Veblen's thought in particular ways, we nevertheless believe that such a reconstruction does him no violence. Veblen's works as a whole, for example, reveal an impressive critique of both status quo–oriented power relationships (social and political as well as economic) and everyday practices (e.g., emulation) that affirm relationships of advantage and disadvantage. They also reveal profound critiques of epistemology (e.g., "habit of thought," the ontology of neoclassical economics) and the social, political, and economic ideologies surrounding the capitalism of his time. Moreover, he accomplishes these critiques through different demystifying methodologies calculated to reveal the ideologically constructed character of institutions (social, political, and economic) in ways quite familiar to critical theorists (e.g., negative dialectics, discourse analysis, praxis, and immanent critique). Broadly, then, Veblen does what critical theorists do in the way that critical theorists do it. The primary difference between Veblen and critical theory, and what we argue should be the basis of a reconstructed critical theory, is his idea that social change is the result of "endogenous evolution" and therefore is (and ought to be) nonteleological. Given the limitations of this work, we cannot deal exhaustively with each of these points. However, we hope that we can deal with them sufficiently to make our point.

Veblen as Critical Theorist

Although the founding members of the Frankfurt school were critical of Veblen,[12] we believe he was in fact within their ranks and that their disagreements with him resulted from their misapprehension of his views, due in some part to his "endogenous evolutionary" ontology and the fact that he was misapprehended by others as a social Darwinist. To substantiate our belief, we first note that Veblen shares critical theory's commitment to the materialization of a society that recognizes the equal dignity, worth, and honor of individual human beings by being continuously responsive to the real and changing interests of its people rather than subservient to dominant interests and ideologies. In *The Theory of the Leisure Class*,[13] for example, Veblen evinces this commitment by affirming that "the test to which all expenditure must be brought is the question whether it serves directly to enhance human life on the whole."[14] The obstacles to this goal include "invidious (ideological) distinctions such as those based on ancestry, ethnicity, gender, race, wealth, caste, class, and employment," since such distinctions involve "a comparison of persons with a view to rating and grading them in respect of relative worth or value."[15] As an example, he points out how the ideological "distinction between exploit and drudgery is an invidious distinction between employments. Those employments which are to be classed as exploit are worthy, honorable, noble; other employments, which do not contain this element of exploit, and especially those which imply subservience or submission, are unworthy, debasing, ignoble."[16] He then points out that any such ideological distinction that "does not enhance human life on the whole" enhances the life of a self-serving and power-conscious elite and "guides the formation of habits of thought" into channels establishing ideologically based ("invidious") standards for judging individuals, their needs and their interests. Such a "process of [invidious] valuation of persons in respect of worth"[17] distorts "the concept of dignity, worth and honor, as applied either to persons or conduct, [and] is of first-rate consequence in the development of classes and of class distinctions."[18]

In a few pithy sentences, then, Veblen places himself firmly within the ranks of critical theory. First, he introduces a critical perspective for the analysis and discussion of all social phenomena, grounding that perspective in a critique of ideology that he considers a distortion of reality designed to camouflage and legitimate unequal power relationships. It is clear from these few passages, for example, that according dignity, worth, and honor to each individual is among the things that enhance human life on the whole and that he considered ideologically based power relationships the central obstacle to be overcome.

Veblen then proceeds to cement his position with the ranks by deconstructing the current "pecuniary culture" in order to arrive at a renewed recognition of the primary importance of noninvidious interests.[19] To accomplish this, he employs three methods of critique familiar to critical theory: immanent critique, praxis, and negative dialectics.[20]

With regard to negative dialectics, "Veblen sought to separate and to polarize elements of American culture" in order to "establish the existence of a dialectical contradiction whose tension and dynamic would provide the promise of radical social transformation."[21] Thus, he constructs "counterfactual histories" (negative dialectics) that were diametrically opposed to prevailing economic theories and everyday preconceptions.[22] For example,[23] in *The Theory of the Leisure Class,* he describes (tongue in cheek) how private property evolved from the theft of women with the purpose of empowering barbarians to pursue leisure activities. This counterfactual idea is set in dialectical contradiction to the accepted ideological view expressed by, e.g., Adam Smith[24] and D.C. North,[25] who explain the emergence and function of private property as the effort to gradually transfer power from the few (e.g., nobles) or the one (e.g., the king) to the many, leading, eventually, to the sovereignty of the human individual.

With regard to immanent critique, Veblen, in his depiction of the functioning of financial markets[26] describes how the internal logic of capitalism brings its ideology into conflict with its practice as shareholders actually destabilize financial markets[27] by forcing business managers to perform shortsighted corporate "sabotage" in the shareholder's (absentee owner's) interest (e.g., downsizing).[28] This is an immanent critique of the mainstream (ideologically derived) "functional theory of finance," according to which financial intermediaries strive to insure that businesses endure and expand.[29] Thus, Veblen regularly employs critical methodologies to bring "habits of thought" (epistemologies) and "theories" (ideologies) into contradiction both with themselves (immanent critique) and in comparison to each other (negative dialectics).

Along these same lines, Veblen critiques both capitalism and Christianity, the dominant economic and religious ideologies of his time, first by bringing them into dialectical opposition and then immanently critiquing each. Regarding the dialogical opposition, Veblen points out the contradiction between an ideology teaching "brotherly love, self-abnegation, and humility," and an ideology promoting the competitive (conflictual) striving for personal pecuniary advantage, or, as he puts it, an ideology promoting "competitive morals" founded in the then dominant ideology of "natural rights."[30] Regarding immanent critique, Veblen points out the contradiction between Christianity's ideological call for humility and the fact that devotional observance is regularly a matter of "prestige and status seeking, as well as other selfish, pecuniary, motives founded in predatory culture."[31]

As a final example, it is worth noting that Veblen immanently critiques the dominant political and social ideology of his time. He points out, for example, that the dominant political ideology resisted the state regulation of private property, although existing property rights were the product of state action (constitutional and statutory provisions). Thus, as far as the dominant ideology was concerned,

> elite use of the state enabled, reinforced, and extended elite domain, considered by them to be part of the productive, if not also natural, order of things. But democratic and pluralizing reforms amounted to intervention and therefore sinful violation of the precepts of laissez-faire and private property.[32]

In all of these cases, Veblen is at pains, as are critical theorists, to indicate how people faced with these internal and external contradictions experience a conflict between their true or noninvidious interests and certain imposed "invidious" interests as reflected, for example, in their emulation of the pecuniary habits of the leisure class.[33] Also, as do most critical theorists, Veblen argues that only a "disintegration" of the current "pecuniary culture" is necessary to a renewed recognition of the primary importance of noninvidious interests.[34]

In addition to negative dialectics and immanent critique, Veblen employs discourse analysis and praxis as critical techniques. Regarding social Darwinism, for example, he points out that "survival-of-the-fittest" discourses are improperly employed in reference to social systems, because such systems "survive" as a "function of structural variables of human construction and [are], moreover, a matter of artificial rather than natural selection . . . [i.e.,] Survival is a matter of human collective choice and power; it is something made, not found."[35] More broadly, Veblen is highly critical "of the transformed use of language: the same words that had been used to describe older legal and economic arrangements were now being used in connection with vastly different arrangements, as if there had been no change."[36] He is very aware that this redefinition was carried out by dominant classes that employed the structures of power to their advantage. Hence, "as Commons presented the matter in his *Legal Foundations of Capitalism,* the central concepts were those of liberty and property, and the central official arbiter of definition and meaning were the courts, especially the Supreme Court," and one of Veblen's major contributions "lies in [his] unmasking and challenging all sorts of linguistic formulations—legal, scientific, religious, philosophic—that are deployed to selectively protect the status quo," especially the status quo of established and specially protected interests.[37]

The role of praxis as a critical method is demonstrated in Veblen through, inter alia, his explications of "idle curiosity," "drift,"[38] and individual purposive action or "workmanship." Each of these involves, in different ways, unforeseen and indeterminate adjustments of theory and practice to both others and the immediate context. For example, it may occur that through "idle curiosity" (perhaps in play), or purposive action (perhaps brought on by "pecuniary exigencies"), or random shifts in the sequences of routinized activities in order to relieve the monotony and tedium of always doing things the same way ("drift"), one may encounter people with different "habits of thought" or find oneself in novel situations.[39] Under such conditions, one may either exit quickly or try various "adjusting" or "coping" behaviors. These "adjustments" are productive of knowledge about how things are interrelated both in the new context or according to the new "habit of thought," and in those with which the person is already familiar. This, in turn, suggests ways of changing accepted habits of thought (theories and epistemologies) and behavior (practice). Things that once seemed important to one another, for example, may no longer seem related, or they may be understood as related in different ways; and phenomena once "unrelated" may now seem important to one another. New problems may come to the fore, either because they were not recognized previously or because behaviors engendered by the new understandings create new problems. As Veblen puts it,

> Any one who is required to change his habits of life and his habitual relations to his fellow-men will feel the discrepancy between the method of life required of him by the newly arisen exigencies, and the traditional scheme of life to which he is accustomed. It is the individuals placed in this position who have the liveliest incentive to reconstruct the received scheme of life and are most readily persuaded to accept new standards.[40]

In brief, the praxis involved in such ambiguous and unpredictable situations requires individuals to exercise ingenuity and produce "unintentional" (i.e., nonideological) change.

In addition to sharing the ultimate goal of critical theory and pursuing that goal with critical methods, Veblen shares critical theory's skepticism regarding "enlightenment thinking," as evinced by his critique of classical and neoclassical economics,[41] and its consequent penchant for interdisciplinary methodologies. Pointing out that in reality we must simply contend with a confusion of "brute facts,"[42] he argues that the difference between theories about what these facts mean "is a difference of spiritual attitude or point of view . . . or in the interest from which the facts are appreciated."[43] More succinctly, "it is mainly a difference in using different frames and preconceptions to organize facts in a meaningful structure."[44] Thus there is, contrary to enlightenment thinking, no epistemological advantage to choosing one theory over another. It is simply a matter of choosing which habits of thought best serve certain interests and goals or which helpfully describe what is going on as interpersonal drift builds up into fluid patterns through emulation, habit, and institutionalization. Moreover, to understand which habits of thought and which goals best serve current interests, Veblen utilizes many different sources and many different types of data, along with the insights and methodologies of anthropology,[45] zoology,[46] ethnography,[47] biology,[48] politics,[49] psychology,[50] and history.[51] And, like many critical theorists today, he employs the findings of both history and science, arguing that "all cultures evolve; generally structures change slowly; sometimes they change rapidly," [52] and that "the exigencies of modern industrial life have enforced the recognition of causal sequence in the practical contact of mankind with their environment."[53]

Thus, both Veblen's goal and approach are those of a critical theorist. However, there is one significant diference. Unlike most critical theorists whose critique involves an "internal or immanent encounter with the existing order, [that] retain[s] a transcendent or utopian [democratic] component," there is nothing in Veblen "comparable to the normative use [of any model] to define or delineate [any proper] direction for economic and social change."[54] Veblen rejects both systemic teleology,[55] as evinced, for example by his rejection of Marx's theory of class struggle,[56] and any form of determinism, whether that of historicists,[57] capitalists,[58] classical economic thinkers (as in the representation of people as hedonistic bundles of desires who merely react to external stimuli),[59] or economic laws based on "timelessness and placelessness."[60] As he puts it, "in all this flux there is no definitively adequate method of life and no definitive or absolutely worthy end of action."[61]

In contradistinction to most critical theorists, then, Veblen conceives of the "critical problem," not as a political problem (e.g., how to establish a radically inclusive and radically responsive democratic structure), but as an evolutionary problem of constructing sequential sets of institutions that prove most effective in coping with the social and natural environment. For him, there is no ultimate, necessary, or ideal institutional structure and process, but only the continual devising of a body of technological and institutional expedients useful for both the extraction of material wealth from nature and the most effective distribution of it (in terms of individual dignity and worth). He thus regards the formation of democratic customs, habits, institutions, attitudes, values, and beliefs as only phases in an evolutionalry process that is without ultimate end or direction. It is upon this moving foundation that Veblen reconstructs critical theory.

Veblen's Reconstruction

Veblen begins his reconstruction with an analysis of classical economics. In both *The Theory of the Business Enterprise* and "Why Is Economics Not an Evolutionary Science?"[62] Veblen objects to the classical notion that human behavior (economic or otherwise) is subject to "laws" or "final causes." Veblen is convinced that laws of economic behavior and other generalizations (about, e.g., hedonistic and utilitarian propensities) are not timeless verities but social and historical constructs meaningful only within the context producing them. For example, Veblen distinguishes his argument from that of Adam Smith precisely on the basis of "Smith's preconception of a normal teleological order of procedure in the natural course."[63] Moreover, by normalizing the "chief causal factor," he affects both the causes and effects in his argument. This, according to Veblen, results in Smith and his successors imposing a "utilitarian philosophy that entered in force and in consummate form about the turn of the century,"[64] creating, not discovering or revealing, "economic man."

Veblen's harshest critique of these utilitarian approaches comes with his statement that "after Adam Smith's day, economics fell into profane hands."[65] Sarcastically, Veblen points out that later scholars did not approach their craft from a "divinely instituted order" or discuss human interests but developed a decidedly metaphysical approach derived from the work of Bentham's simple "theory of value."[66] Later, he demonstrates how the cause-and-effect approach taken by these scholars lacks explanatory power, particularly when dealing with the intermediate concepts including "sensuous impressions and the details of conduct."[67] Veblen defines these sensuous impressions in the context of hedonistic wants, placing them in his economic language. The discussion moves beyond one of utility, pointing to the inadequacy of the utilitarian conceptual framework.

In contrast, Veblen offers a historical and "evolutionary" economics without trend or consummation. Economic behavior is explicated in terms of the day-by-day adaptation of economic institutions to actual events of immediate importance to extant individuals, and economic change is elucidated as a process of choosing the form of economic institution "most fit" to current historical happenstance. That is, economic behavior is accounted for via the methods people developed for dealing with the problem of wresting a living from the historically constructed reality presented to them. Economic institutions are nothing more than habits and tacit agreements about how to proceed in that wresting from the current environment (social, political, and natural).

A key tenet of critical theory is that "positivist" theories and methodologies are distorted because they ignore the internal human history of the social world and its institutions. Economic institutions and process, for example, are understood by critical theorists as expressions of human of interests, needs, and psychodynamics rather than as the result of "natural laws" or "invisible hands" working themselves out through human beings and their institutions.[68] Moreover, because the institutions are socially constructed, they may be called into question legitimately and changed to meet human interests in ways not necessarily apparent in any objective account of the current situation.[69] Therefore, both Veblen and critical theory share an attack upon mainstream, positivist, and instrumental modes of thought. All are considered distortions of reality in that they fail to emphasize human needs and purposes (intentionality) over behind-the-scene natural laws or system needs and values.

Veblen's *Theory of the Leisure Class* develops and deepens his contribution to critical theory.[70] The theory contends that certain unintended ills arise out of the habits and tacit agreements crafted to facilitate the wresting of a living from the current environment. First, the agreements usually privilege individuals. Second, the leisure class, in pursuit of these privileges, works to inhibit the "natural selection" of institutional forms better suited to the historically emerging social and economic environment. This pursuit is reinforced by both the propensities for emulation by other classes and the habitual patterns of thought ingrained by current institutional arrangements. Thus, the leading industrialists of Veblen's time were not the fittest agents of evolutionary advancement but parasites on the labor, creativity, and innovation of others. The extant price system, by concentrating resources in business owners rather than in those directly involved in production and innovation, impeded the advancement of industry and the industrial arts required for the next step in social and economic evolution. Similarly, rents were a ransom extracted from the productive population, sabotaging innovation in the process.

The Theory of the Leisure Class adds depth to Veblen's contributions to critical theory by stressing his idea of "latent functions." For example, while art may serve manifestly to brighten a room, it also serves to signal good taste or sophistication (its "latent function"). Conduct generally, and patterns of consumption particularly, can be meaningfully understood only in terms of both latent and manifest functions, and the enhancement of status (power) is preeminent among those functions. So, for example, ostentatious consumption, philanthropy, great expenditures in showing others a good time, and cultivation of the arts and literature all serve to establish and enhance status, eliciting deference that magnifies the power of those engaged in the activity.

All this is clearly an early recognition of the latent oppressive functions of discourse, religion, institutional structure, and social process emphasized by recognized critical theorists. Critical theorists, for example, pursue ideological critique as a negation of the latent content in the language, assumptions, axioms, and ontology of the dominant ideology. They contend that this latent content holds power over people and causes a crisis in legitimacy that propels the development of new (less oppressive and more emancipating) ways of getting things done. The critical difference, as we shall see, is that while critical theory's analysis either leads to philosophical cul-de-sacs (e.g., Adorno and Marcuse) or remains largely abstract or theological (e.g., Habermas and Horkheimer), Veblen actually connects language, meaning, axioms, ontologies, discourse, and immanent critique to the contexts and situations of lived action, social relationship, and the circumstances of perpetual change.

Veblen's *Theory of the Business Enterprise,* "Why Is Economics Not an Evolutionary Science?" and *The Theory of the Leisure Class* all anticipate the methodology of critical theory. The methodological basis for critical theory's immanent critique of society is the dialectical logic of Hegel.[71] According to this logic, truth lies in the process of critique (including self-critique), and positive change lies in the process of challenging established reality. This continuing critique is meant to unfold the potentials in a given reality and to actualize them in practice.[72] This critique and challenge are accomplished through both discourse and praxis. The idea is that both theory and analysis become less and less distorted the more freely, the more broadly, and the more often diverse ideas, worldviews, and reactions are communicated and discussed.

While Veblen does not accept the teleological aspects of Hegel's theory, he does employ Hegel's dialectical method. More precisely, following Veblen, action becomes a form of discourse through workmanship, curiosity, and emulation. The false consciousness engendering emulation can be overcome through both curiosity about the unique and the human propensity for activity tailored to the efficient achievement of goals satisfying genuine interests. This idea is precisely reflected in critical theory's contention that praxis is the key factor differentiating critical theory and analysis from both ahistorical "fact gathering" and the cataloguing of abstractions.[73] However, unlike current critical theory, "praxis" here refers to concrete, conscious, practical action constituting a critique of the current situation. This praxis should eventually lead to direct action against the extant economic system. Thus, in critical theory itself, this conflation of discourse and action is understood and appreciated but not given the pride of place accorded it by Veblen.

Throughout Veblen's economic thought is a deep appreciation for the synergy among forms or modes of living, individual patterns of thought, institutional structure, and community organization. Schemes of thought, as he would put it, are reverberations of life schemes. Therefore, as an example, one's position in the economic sphere determines one's habits of thought. All habits (ways of acting, ways of thinking) grow within groups as people seek the fittest institutions for the situation. Concerning this synergy, Veblen finds instances of maladaptation particularly interesting. When habits of thought lack congruity with the historical economic setting (with, e.g., the occupational structure or available technologies), distortions occur in the evolutionary process. Some distortion is "natural" because institutions are adapted to prior stages in the evolutionary process and thus never fully accord with the immediate situation. Some distortion, however, is "sociological." It is occasioned by vested interests fending off change, the persistence of inappropriate patterns of thought, and the attractiveness of occluding metaphors.

Of primary concern to critical theory is the tendency of "systems" (e.g., capitalist economic systems) to dominate the "lifeworld" of individuals.[74] The lifeworld is where people acquire shared meanings through everyday communication and the practices of everyday life. In *The Theory of Business Enterprise*,[75] Veblen addresses this concern as he argues that people interact with the natural world to satisfy material and social needs. In the process, they create a culture that "adds to" nature and reconstructs reality to meet human needs and interests. When a system dominates this world, people are treated as objects, not subjects. This results in a "standardization of the workman's intellectual life in terms of mechanical processes."[76] People, in effect, become "capital" to be manipulated according to the system's interests. They lose their animus, their individuality, for a "standardized quantitative precision."[77]

The system and the psychology it engenders (i.e., "false consciousness") through this mechanistic discipline, ultimately act "to disintegrate the institutional heritage, of all degrees of antiquity and authenticity—whether it be the institutions that embody the principles of natural liberty or those that comprise the residue of more archaic principles of conduct still current in civilized life."[78] This disintegration illustrates how humans, acting within their institutions, can unmake them. They can employ critique and self-criticism as a dynamic to facilitate an awareness of both their status as objects and their eminence as self-subjugators, participating in the system according to its directives and in pursuit of system interests. Through the dynamic of this destructive process, people can craft an antithetical system that reveals how such processes limit human interests.

Furthermore, Veblen seeks to counter this same domination of his own thought by viewing the economic process without imposing on it any other categories than those it created itself through its growth, development, and failures. As stated in the preface to *The Theory of Business Enterprise*,

> In respect of its point of departure, the following inquiry into the nature, causes, utility, and further drift of business enterprise differs from other discussions of the same general range of facts. Any unfamiliar conclusions are owed to this choice of a point of view, rather than to any peculiarity in the facts, articles of theory, or method of argument employed. The point of view is that given by the businessman's work, the aims, motives, and means that condition current business traffic. This choice of a point of view is itself given by the current economic situation, in that the situation plainly is primarily a business situation.[79]

Veblen seeks this point of view to emancipate himself from the "cultural absolutism" of mainstream economists who assume that the categories of the present economic system exist eternally and continue unchanged.[80] In brief, both Veblen and modern critical theory seek emancipation through insights gained by a critical awareness of the categories of thought employed in analysis and theory and by a critical awareness of the needs and interests of the self and others who are actually engaged in the activity under consideration. Such awareness is thought to facilitate a transformed consciousness or "perspective transformation" from which people may fashion an alternative.[81]

A Veblenian Solution to the Theory-Practice Gap

Most important to critical theorists is the reconstruction of theory as lived experience. Theory is only meaningful, in their view, if it is tested in historical struggles for emancipation. The validity of any alternative to negated ontologies, axiologies, and epistemologies lies in its actualization by people in their day-to-day activities. By this criterion, critical theory has failed. There seems to be no evidence that any immanent critique currently propels or guides the action of any movement, party, interest group, or regime. Moreover, forms of life that are distinctly nonemancipatory thrive in many places, even within societies dedicated to an ongoing self-criticism of their institutions and regimes. Fundamentalist Christian movements in the Unites States come to mind. The people involved seek definitive, unconditional guidance from an authoritarian source, embrace a distinct teleology, and are steadfastly unimpressed by immanent critiques of Christianity. Worldwide, people seem happy with a range of more or less emancipatory political institutions, and even where the institutions are ostensibly structured for self-criticism people may employ them otherwise. Elazar's[82] "traditionalistic" political culture still obtains in certain parts of the United States, and his "moralistic" political culture is not necessarily emancipatory as it comprehends the individual and the group as integral. Ironically, Elazar identified "individualistic" political cultures, generally held to be on the rise with increasing urbanization, as the most corrupt. In brief, nonemancipatory discourses, institutions, and processes that do not respect the integrity of the individual (in the sense of self-rule) seem to be working, and nowhere do we find a working model of critical theory as delineated by modern critical theorists.

The roots of this failure lie in one of critical theory's most profound premises and in one of its most tacit assumptions. Critical theory's premises are rooted in a thoroughgoing and radical form of social construction. While the original Frankfurt school grounded itself in the norms and objective physical dynamics of Marxist materialism, this changed as critical theorists relentlessly employed both their thoroughgoing constructionism and their analytical methods in critique of extant and developing political, economic, and social orders (e.g., fascism, capitalism). Subjected to these methods and premises, all assumptions of any existing order became social constructs, as did all epistemologies. More important, all motivating forces and values became mere projections (à la Freud) of categories and habits of thought engendered by situation, context, and perhaps shared cultural neuroses. Necessarily, the Frankfurt school's own grounding of critical theory in transcendent historical and value categories became suspect as well. Such categories as "class" and "the ideal speech situation," such values as "emancipation" and "liberation," and such epistemologies as "dialectical materialism," must have been historically constructed and so begged for immanent critique (after all, "grounded in transcendent categories" is something of an oxymoron, and "grounded in transcendent value categories" is doubly so).

This sweeping social construction permanently unhooked praxis from theory. No more thorough an "ungrounding" could occur, for example, than the explicit denial by the Frankfurt school in the 1940s of any claim to having uncovered any sort of truths in aid of revolution or social reconstruction. Truths, they argued, could only be uncovered through the struggle for emancipation. From social interaction alone might emerge some shared truths, and only from personal action might a self-understanding of truth arise.[83] From a praxis standpoint, this stance seems oddly and paradoxically conservative. It seems to suggest that neither revolution nor reconstruction can be organized, managed, or led. Each must simply occur (incrementally?) as we go along, apparently in pursuit of values already present in the overthrown or unreconstructed order. Moreover, if we take this denial of truth seriously, how do we know if there is a theory-practice gap? On the one hand, how can praxis be out of step with a theory disdaining all vision, direction, or valuation of certain behaviors over others? On the other hand, is there actually any theory to be out of practice with?

Continuing in this radical constructivist vein, Adorno and Horkheimer argue that there is no available form of linguistic expression that does not accommodate itself to the dominant habits of thought.[84] This is certainly consistent with critical theory's scrupulous social construction, but where do we go from there? How do we construct a verbal critique that is both understood and not inherently a tool of the system it is seeking to change? Without such a critique, how do we close the theory-practice gap? Perhaps, again, through action; but then we are returned to the quandary of having no truths to guide and evaluate that action. The same problem arises regarding the Frankfurt school's theory of the "totally administered society."[85] If new forms of technology, new modes of organizing, new configurations of class, and new forms of social control have extended capitalist control over all aspects of life and secured political, social, and cultural conformity, we are trapped in a philosophical and methodological cul-de-sac.

In brief, if all categories for comprehension, all systems of thought and discourse, all methodologies, and all institutions are socially and historically constructed, controlled, and conformed, there is no way out.[86] We are adrift without normative or methodological stars, and the theory-practice gap cannot be closed. From this impasse, it has proven a short step to radical relativism, pessimism, and nihilism. If, for example, no concepts of justice, equality, emancipation, tolerance, and happiness are grounded in anything that guides their social construction, they are hardly reliable guides to practice. In fact, the mere idea that justice and freedom are better than injustice (which may, after all, be constructed easily as, say, "a discriminant allocation of resources") and oppression ("a guiding hand keeping us securely on course") gains no purchase.

The other problem lies in critical theory's tacit assumption that "human integrity" and freedom must be realized together. One problem with this assumption is the paradoxical nature of what we understand as "freedom." The idea that freedom includes not only an absence of interference but also an effective power to act[87] suggests that forgoing a certain amount of self-rule may be necessary to organize effectively in order to accomplish things that one cannot do alone. Therefore, both allowing interference with personal choices and allowing others to make certain choices for one may sometimes be necessary to be free.

Habermas answers these concerns by grounding concepts of justice, equality, and emancipation in the intersubjective construct of language. His idea is that social norms may be legitimated if they either are achieved or could have been achieved through reasoned discussion under idealized conditions. The idealized condition arises from a mix of three interests that all people share: an interest in effectively dealing with the natural world (a positive, scientific, and technological interest), an interest in effective interpersonal communication and coordination (a hermeneutical interest), and an interest in increasing power (an emancipatory interest). Satisfying these interests for everyone requires a discourse unconstrained by authority, tradition, and metaphor. No participant, thought pattern, or methodology should be privileged, no assertion should be free from challenge, and the participants should be truthful. As this ideal speech situation provides equal power to all and prevents decisions based on greed, dogma, tradition, or coercion, it should produce undistorted agreements upon what is just, equal, and emancipatory. Eventually, this idea culminated in the notion of "discourse ethics" and certain "rules of reason" that "sought to reconstruct the intuitive grasp of the normative presuppositions of social interaction possessed by competent social actors generally."[88] The problem is that the practical problems of striving for the ideal speech situation are immense.[89] Consequently, it serves not as a pragmatic way out of the theory-practice gap but as an abstract transcendental standard that is unrealizable in practice.

There is a second problem. For Habermas, the general conditions of the ideal speech situation and the rules of reason describe a particularly responsive form of plural democracy. The idea is that a diversity of communities and participants, following the same set of rules regarding discourse, may establish diverse, flexible sets of norms as legitimate for a given community, but not all communities, and thus avoid all forms of authoritarianism and privilege. But it would seem that this can happen only if the institutions effectuating and mediating discourse (e.g., executives, legislatures, courts, and bureaucracies) not only attend to the manifest values of the system, but engage in a search for both latent injustices and latent values. To accomplish this, they must concern themselves with both particular outcomes and the systemic, society-wide patterns of effects implied by alternative policies. It also requires that they be very flexible in their procedures, policies, and possible solutions to ensure an effective response to the latent oppression and injustices they find. Given that values and effects are always multiple, interdependent, potentially conflicting, and often (empirically) irreconcilable, there is a real risk that decision making will degenerate into an ad hoc balancing of competing goals and interests. The fitness an institution gains by having a defined jurisdiction, a distinct purpose, and an established set of rules and procedures is lost as it becomes more outcome-oriented. It takes within its purview the total society-wide impact of its decisions and so becomes more flexible in its procedures and policies. With declining "fitness," in this sense, the theory-practice gap widens.

An intriguing suggestion for closing the theory-practice gap left open from the Frankfurt school through Habermas may be gleaned from the writings of King and Zanetti.[90] King points out that "practical theory happens at the nexus of the real and the ideal . . . at the in-between of scholarship and practice."[91] For King, "making meaningful theory is about judging, valuing, intuiting and making choices" at this nexus of positivism (with its generalized formal theories) and the lived reality of human beings with singular needs and interests. Zanetti suggests that the gap may be closed at this nexus by participatory research. Her idea is to develop practical knowledge, "hegemonic consciousness," and political action by establishing a "democratic tension" between the formal knowledge of the academic researcher and the popular knowledge, personal experiences, feelings, and spiritual expressions of particular communities. Pursuant to her approach, the theory-practice gap would close as the research subjects (i.e., the community) democratically determined research objectives, research methods, and research objects and then democratically adjusted the research process as they went along in order to ensure the realization of knowledge directly useful to them. In this way, employed methodologies would be refined, refocused, and even altered drastically through an emergent process of collaboration and dialogue that might be expected to inform, motivate, and increase the self-esteem of community members while developing community solidarity.[92] Thus, self-sufficient, self-assertive, self-determining communities would arise through an integration of theory and practice.

Zanetti and King are probably correct. As Baldwin[93] suggests, if we limit the domain of theory construction in some useful way (say, to a set of dimensions of our current reality that are interesting or important to us given our current needs, interests, and goals), we can evade certain problems of theory choice arising from the fact that all more inclusive theories are equally unverifiable, equally unfalsifiable, and equally unlikely to increase in verisimilitude across domains.[94] Conversely, if we choose and develop theories based upon their relevance to community concerns and their efficacy in resolving immediate problems, then the objectives, values, and interests motivating the theories (and implicit in their development) can help identify their domain.[95]

In this context, it is important to recognize that predicating theory development and choice on information from a subset of all available information, deliberately restricting theory choice and development to particular problems, and adjusting methodologies throughout the entire research process in order to address particular needs and interests seems a form of instrumentalism. Certainly, instrumentalism can lead to an independent role for values in the theoretical process. As problem selection is no longer implied necessarily by the selection or dominance of a received theory, the "best" theoretical construct can depend upon the choice of problem, interest, or need as informed by values and shared experience. Nevertheless, instrumentalism still has more interesting implications for the role of values, interests, and needs. Because an instrumentalist approach does not define the problem from which researchers must choose, nothing prevents necessarily the ever-increasing proliferation of problem areas, each with its own theoretical structures, paradigms, and methodologies that can address particular needs and interests. Indeed, a "pure" instrumentalist perspective fosters this proliferation. After all, it is the efficacy of a theory or a methodology within its domain that counts, not its consistency across domains.

The only question remaining concerns the defining of relevant problem areas. How does a community recognize, identify, and frame its needs and interests? Does one allow for an ad infinitum division into subproblems? Does the status of each subproblem (or set of subproblems) equal the status of the (relatively) more general problem? Without a technique for problem definition, the research and theory choice process either cannot begin or becomes stuck forever at its first discursive interaction among the individuals or groups involved. If one replies, "we will democratically decide," what does that entail other than a Habermasian ideal speech situation with all its attendant difficulties?

Obviously, community values, fundamental conceptualizations, and traditional categories of thought (i.e., the community's ideology) could play a significant role at this point. As Abel and Oppenheimer point out,

In a situation where problem "b" is explained most accurately by theory "B," while another theory, "A," best explains a subset of "b" (call that subset "a"), instrumentalism would direct the choice of "A" if you're interested in "a" and "B" if you're interested in "b." But should the major interest be in "a" with a residual interest in "b," "A" would be chosen even though it is not as good a theory as "B" in dealing with the problem as a whole. Again, if the interest is in "a," but "A" requires a reordering of priorities or a radical alteration of fundamental concepts, where "B" does not, "B" might well be chosen though it predicts less well. In Beardsley's terminology, a community's anchoring point (what it is willing to consider as subject to inquiry and change) or its ideology and moral bias circumscribe any tendency toward "rampant instrumentalism."[96]

At the same time, as instrumentalism is a game played with, and to a large extent governed by, data from a limited domain, one may remain a "true believer" in a theory or a methodology only so long as there is a domain within which the theory or methodology instrumentally dominates its rivals. In addition, the domain must be sufficiently important to justify the theory or methodology based on the domain. In brief, a "reflexive" relationship between instrumentalism and ideology keeps each from running amok.[97]

Theories and methodologies, then, are discardable when they are no longer defensible ideologically or instrumentally in a domain. Note that this domain need not be "practical," but can be of purely theoretical interest (e.g., the interface of two theories predicting quite different things within one arena and hence establishing expectations of a "critical experiment"). Consequently, theories may coexist without generating the need for or interest in a "critical experiment." In brief, this interplay of ideology and instrumentalism emancipates scholars, researchers, communities, and individuals from the requirements of strict empiricism (which are impossible to fulfill and insensitive to the social and cultural dynamics within which they function). On the other extreme, it liberates them from the dictates of pure ideology, which seeks to create reality in terms of its own fundamental premises.[98]

Habermas, King, and Zanetti seem to have exactly this interplay of ideology and instrumentalism in mind. Habermas, for example, makes no apology for his confidence that only by employing something close to his ideal speech situation might a radical form of participatory democracy (or at least some synthesis of liberalism and civil republicanism arise).[99] King is equally committed to a "more democratic . . . scholarship where knowledge and understanding are emergent," and everyone is engaged in a "participatory collusion."[100] Her suggestions for "healing" the theory-practice gap are instrumental to that end and intended to empower marginal voices. Finally, the entire point of Zanetti's proposed methodology is to democratize and sociopolitically empower everyone and every community by overcoming the conflicts engendered by principles challenging the moral framework of complete democratic inclusion. In brief, the ingenious suggestions for closing the theory-practice gap offered by Habermas, King, and Zanetti are meant to be instrumental in achieving the same transcendental ideological vision characteristic of critical theory from the Frankfurt school on.

From a Veblenian perspective, however, the idea that democratic participation of any sort (regardless of how accomplished) necessarily emancipates or sets us upon the proper path to an awareness of possibilities as yet unrealized is ultimately an act of faith. There is no reason for supposing a priori or a posteriori that any particular vision, path, or practice is, was, or will be desirable at all times and in all contexts. Multiple community directions, forms, shared meanings, values, and visions emerge simultaneously and persevere together (more or less well) over time. They cannot be effectively teased out, and there is nothing but this multiplicity and variation to reveal. In fact, Veblen might find it telling that neither the vision nor the proposed methodologies of Habermas, King, and Zanetti were communally derived. For him, this could constitute a sign that from the community's perspective no significant interest or need is addressed by them, or that the needs are addressed sufficiently in other ways, or that the community is deteriorating.

Although Veblen eschews all transcendental ideologies and methodologies, there is little worry of generating radical relativism, pessimism, or nihilism. This is primarily because Veblen does not consider it sufficient to deconstruct in the hope either that an alternative is struggling to emerge, or that immanent critique will spur the creativity necessary to construct a more emancipatory way of getting things done, or that a transcendental dynamic is operative. Rather, his idea is to unhitch one's thinking from dead metaphors or inapplicable patterns of thought that occlude our vision of what is actually going on. Critique should be pursued in ways that reanimate thought and behavior with new patterns of thought that are more appropriate to realizing human interests in the current context. It is not to be directed necessarily against or toward establishing any particular sociopolitical form, nor pursued necessarily by any preferred means or methodology. Thus, Veblen does not merely critique classical economic thought but takes a direct look at exactly what is actually going on as people interact and discourse economically in his time. Given what he sees, he propounds an "evolutionary" approach anticipating change in the things he sees and deriding the tendency of classical economists to interpret through current categories what they see historically. In the process, he implies an Evolutionary Critical Theory without the ideological elements, the tacit assumptions, or the problem of ungrounded norms encountered by critical theory in its present incarnation. It is from this Evolutionary Critical Theory that a solution to the theory-practice gap is gleaned.

Veblen's idea is that that the gap is largely illusory. There never really are such things, at least not for long. This is simply because thought and theory develop as people deal with nature, their biological needs, and extant human constructs. These thoughts and theories (including thoughts and theories about what current needs and interests are) "evolve" in the same sense as that term is used to subsume observed changes in organisms as they respond to biological needs given their situation, context, and history. There is no teleology—there is simply a pragmatic, prolific, undirected trial and error. So long as this process is unconstrained artificially (e.g., by human institutions), everything that can happen will happen, and all that can find purchase will remain (at least for now).

Habits of thought, then, will proliferate and either adapt or fail. Should they fail, people, institutions, and civilizations will develop new ones, suffer and muddle through, or die. Thus, for example, Veblen concluded that without new habits of thought (e.g., Keynesianism?), the capitalism of his time (a set of suboptimum habits of thought) would evolve into either totalitarianism or socialism, neither of which was desirable. Under totalitarianism, vested interests would strengthen and direct production to the benefit of the few. Under socialism, unions would form and hinder production to keep wages high. In either case, free and productive experimentation would be hindered as institutions were sent off into less productive evolutionary dead ends. In any case, the necessary alignment of thought, theory, and situation closes the theory-practice gap.

To understand exactly how this works, consider what must obtain to have a theory-practice gap. First, there may be a gap in the sense of theory constituting an abstract ideal never realizable in practice (i.e., the Socratic and Aristotelian sense of "theory"). Veblen argues that this can never obtain simply because there can, in fact, be no passive knowing—no moving to a place outside of time, context, and situation. Rather, such a habit of thought must arise to some purpose and must serve some human interest. Idle curiosity would be the Veblenian interest that springs to mind, but it seems that such habits of thought also provide some comfort and some satisfaction in making sense of apparently chaotic experiences. Such habits of thought go in and out of vogue, of course, perhaps depending upon the challenge new experiences offer to accepted paradigms, perhaps depending upon how much leisure there is to indulge idle speculation. In any event, there is no theory-practice gap. What appears as such a gap is actually a clash between the habits of mind and behavior that these theories engender and other theories and habits of thought brought about by other experiences, perhaps more vital to the context. People must then decide which, if any, of these more vital interests to give up because of the interest in abstract theorizing. Making such choices will cause the popularity of abstract theorizing to wax and wane; but as everything that can happen will happen, and as the human interests theory-practice gap serves are likely to remain, it is unlikely to ever disappear.

Next, there may be a theory-practice gap in the sense that the theory would produce habits of thought and action better serving contextual and situational interests if only people would act according to the theory (i.e., "the spirit is willing but the flesh is weak"). Veblen's idea here is that as change is always evolutionary it is often (though not necessarily) incremental. New theories are often "go-betweens," conjoining old habits of thought and behavior with new habits and facts in ways that insure continuity. This serves easily identifiable human interests in instrumental control, directing devices, effective communication, and the eventual transformation of consciousness necessary to evolve. Thus, again, there is no gap but, at best, something of a lag in perfectly fitting theory to practice, and again, the flesh will become willing, suffer and muddle through, or die. In any case, the theory proves itself out in practice, negatively or positively.

Third, there may be a gap in the sense that while a theory is the "best in situation and context," the people are ignorant of the facts and experiences necessary to realize that this is true. Additionally, things may work tolerably well under the "less best theory." Now, from a Veblenian perspective, how could this be the case? Should people lack the facts and experiences (and so the thoughts) indicating that current theory does not work well, theory and practice must be well joined.

Finally, there may be a theory-practice gap in the sense that people create and support an ideal divorced from their actual practice and experience, choosing to behave as though the ideal were true. Generally, such an ideal is mythological—that is, not so much abstract as fantastic. It is not simply that the theory is contrary to experience, but that experience is denied. People ardently imagine the world to be working one way and persistently ignore, explain, or imagine away contrary experiences and indicators. All a priori belief systems (Christianity, Marxism, capitalism, democracy) have theory-practice gaps in this sense, and they all argue from design. God or nature has a "final cause" for the world and either a design or a set of latent templates operates in the universe to organize its accomplishment, though we may not be able to grasp either the end or the process fully, if at all—consider, for example, God's nonanswer to Job.

Of course, Veblen's evolutionary perspective displaces design, and to his mind, theories from design can produce no "proper action." As with Socratic theory, however, such theories must serve some human interest or they would not arise, and both the intensity and the resilience of the dedication to such theories indicates that some very important interests must be served for at least some members of the community. Again, the satisfaction of these interests must struggle against contending theories useful in the satisfaction of other interests. In practice, people will choose and prioritize, and because everything that can happen will happen, we are likely to see a hodgepodge of adaptive attempts happening all at once. Those adopting theories that fit experience will flourish through efficacious practice. Those persisting in their dedication to a theory divorced from practice will suffer, muddle through, or die. Some will "imagine" harmony between their theory and what is going on around them, perhaps acting effectively as a result. Others will simply not relate their theory to their practice in any meaningful way, but simply hold the two distinct and perhaps act effectively as well. In these last two instances, of course, either theory and practice are reinterpreted in light of each other or a separate theory tacitly guides practice in certain realms of the individual's endeavor, allowing the "mystical" theory to serve the individual's interests in other realms. In any case, either theory and practice harmonize eventually, or society and its habits of thought suffer, may at best muddle through, and ultimately must die.

Some current critical theorists endorse the idea that the theory-practice gap is largely illusory. King, for example, holds that "the theory-practice gap is not real; it is a construction of our making and can be deconstructed in a meaningful way."[101] The difference is that while King understands "breaking down the walls that separate scholarship and practice . . . [and] dismantling the belief systems that privilege some over others"[102] as necessarily adaptive, Veblen would argue that walls and privileges serve human adaptive interests well under some circumstances. Properly contextualized, they may even serve some of King's interests in seeking to dismantle them. That is, they may be the "fittest" way to live with the "contradictions of the paradoxes implicit in the relationship" between theory and practice.[103] In this sense, Veblen's evolutionary critical theory casts current critical theory and its breakdown of all dualisms in much the same light as Dennard casts the whole of postmodernism for him.[104] It is a transitory form of critique marking a transitional period in an evolutionary trend. It is not the next stage in social, political, academic, intellectual, or human development.

In brief, Evolutionary Critical Theory closes the theory-practice gap by assuming no teleology or determinism of any kind and holding no preconceptions of what priorities, privileges, or absences thereof serve best. It does not assume, for example, that human interests, needs, and psychodynamics are always (or never) served best by promoting "human integrity" and "individual freedom" (together, separately, or in some priority to one another). It also assumes no linearity, colinearity, or multilinearity. Apparently "next" stages in economic or social development may be skipped. Regression and stagnation are both likely, and each is good or bad (more or less) only (but definitively) in context. The key point is that people do (at least eventually) adopt those institutions, privileges, or lack thereof that fit the circumstances best despite the concept of "best" institutionalized in discourse (including critical discourse) and the dominant ideology. If for some reason they do not (because of, say, external repression or the ravages of disease or their inability to shake inapplicable habits of thought and behavior), they suffer and eventually die. Thus, in response to the proposition that "democracy is a very bad form of government except for all the rest," we suspect that Veblen might at best rejoin, "Well, maybe right now." That is to say, most theories of change seek or assume an "inner dynamic" in the sense of sources or directions of transformation that are built into the structure of phenomena. Veblen resists this quest because it tempts people to read their preconceptions into history and to attribute necessity to what is actually chance, whim, human creativity, and other conditional factors. Simultaneously, Veblen eschews the tendency of pessimists, relativists, and nihilists to focus on the historically contingent to the exclusion of immanent propensities. In contradistinction to both approaches, Veblen seeks to grasp apparent happenstance as a reverberation of the character of the social system under construction as people seek the institutions fittest to their historical circumstance. It is in this sense that he speaks of change as a result of "external" influences and the "evolutionary push" with a "sense of direction."

Evolutionary Critical Theory, then, is a theory of constraint and response identifying *potentials* for change in the current situation that are created by people themselves as they go about meeting their interests on a daily basis. It identifies immanent contradictions, maladaptations, opportunities, internal human expectations, and more or less successfully emergent adaptations that do not determine what will happen as they are conditioned by countervailing conditions, hostile interests, and inappropriate modes of thought and action. Every "new stage" is thus a hypothesis that this configuration of institutions, thought patterns, discourses, and patterns of social interaction is the fittest for dealing with the limitations of the prior stage as revealed in its contradictions.

Now, given that Evolutionary Critical Theory is not teleological or deterministic, some prior stage or less emancipatory configuration of discourse and institutionalization might, so far as it is concerned, constitute a good one in the current situation. That is, regression or repression may be adaptive. Whether they are would depend upon what is revealed by a close assessment of the actual contradictions, maladjustments, resources, and immanent human interests at play. Most starkly, repressive, nondemocratic, nonemancipatory systems violating human integrity are natural evolutionary responses to some contexts. In those contexts, they may constitute the fittest choice of institutional structure. What we currently consider "advanced" discourse and institutional structure should not be read onto either the past or the future. Democratic processes emphasizing human dignity and individual liberty (emancipatory systems) may not always be the fittest, the most adaptive, or the most stable.

In Evolutionary Critical Theory, then, emancipation, equality, and human integrity lose their transcendent (ideological) status and gain whatever meaning and priority in human behavior that the actual context provides. Their desirability (not just their achievement) becomes historically contingent, depending upon the urgency people give them and the resources people can and are willing to dedicate. Evolutionary Critical Theory, then, closes the theory-practice gap. It maintains that societies either die (more or less quickly) or eventually adopt institutions that fit the circumstances best, and that death or adaptation is guided by an immanent critique of both institutionalized discourse and the dominant ideology that is achieved through the behavior of the people directly involved. The reason there seems to be no evidence that any immanent critique currently propels or guides the action of any movement, party, interest group, or regime is that critical theorists are looking in the wrong places. They should be looking for the directing immanent critique within the actual behavior and language games operative within the forms of life extant and developing in a society. They should not be abstractly critiquing (however immanently) the dominant ideology.

Moreover, it is not surprising that distinctly nonemancipatory forms of life may thrive in many places. The configuration of discourse, institutions, and ideology in any place will depend upon the conflicting immanent interests both among people and within the same person. It will also depend upon the resources, the priorities, and the will of the people involved. Those interests may, in their estimation, be better met by something other than a highly sensitive and responsive democracy.

Conclusion

We end this discussion with the following points. Veblen's contribution to critical theory affords a nonteleological, nondeterministic evolutionary social constructionism and concomitant analytical method that makes sense of social and historical situations tending to defy other scholarly efforts at explanation. Evolutionary Critical Theory demonstrates how in practice a socially constructed process of adaptation does not always move toward a more democratic, more emancipatory situation. Using the conception of "fittest" solutions, we can understand how, despite our best efforts, we can continually discover examples of social life antithetical to the ideals presented by other critical theorists.

Simply demonstrating that an Evolutionary Critical Theory can reasonably and effectively exist is not enough. To make this contribution both useful and practical, we must then demonstrate how it can illuminate the intersubjective experience of good governance at the core of the discipline of public administration; how it provides for the synergy among hermeneutics, institutionalism, and traditional social science; and how it can effectively bridge the theory-praxis gap while creating a space for "endogenous evolution." Therefore, we must begin to systematically address some of the major issues faced by critical theorists as well as public administration scholars and practitioners. By using the lens of Evolutionary Critical Theory, we now can understand, reconcile, and possibly even internalize many of the roles of power, emancipation, and the "good society." Thus, if we can understand the nature of power in these intersubjective experiences, we can then understand how institutions, organizations, and individuals interact. This in turn drives us to pursue how power is understood in this context, and how public administration plays an integral part in our understanding of its roles in society beyond simple notions of domination or social control.

6

Evolutionary Critical Theory, Power, and Emancipation

After establishing the idea of a nonteleological, Evolutionary Critical Theory, capable both of being a synergistic agent and of endogenous evolution, we see where it can address many of the concerns of critical theorists in the tradition of the Frankfurt school. Now, we begin to focus on the major issues central to the study of public administration and critical theory. To this end, we intend to uncover the power relations intrinsic to social, political, and economic practice. In the process, we challenge the conventional view proffered by critical theorists who argue that bureaus and agencies are necessarily hostile to human freedom and self-determination, hindering the development of open, intersubjective experiences central to their idea of emancipation. Many critical theorists and theorists in public administration generally argue that since bureaus necessarily take on self-sustaining, self-directing lives of their own, they can evolve values and goals contrary to the actual interests of those they govern and ineluctably dominate the individual's lifeworld. To defeat this natural tendency, critical theorists have maintained that public agencies must either be captured as social (as opposed to political) institutions or contained within a "minimalist administrative state" that is responsive to a radically democratic public sphere operating according to the principles of the ideal speech situation. The current progress of the administrative state worldwide thus leaves them in despair.

We argue, instead, that Evolutionary Critical Theory, while sharing the goals, methods, and critiques of traditional critical theory, allows for a nonteleological and nondeterministic (endogenous) social evolution. Such a social evolution then allows us to posit a more dynamic social constructionism than that of either the Frankfurt theorists or their Habermasian heirs. In particular, Evolutionary Critical Theory denies that highly responsive forms of democratic pluralism necessarily emancipate and that the best institutional arrangements are necessarily free of power relationships.[1] Rather, we contend that it is these same power relationships, considered the bane of emancipation, that arguably can become the ontologically driven basis for developing, maintaining, and guiding the intersubjective experience of good governance existing at public administration's core.

In addition, the traditional understanding of critical theory, that an inherent contradiction exists between bureaucracy and emancipation and between agencies and agency, stems both from its oversensitivity to the invidious aspects of public organizations and from its overtotalizing estimation of their actual and potential impact. This insensitivity to the indeterministic and (noninvidious) emancipating characteristics of public organizations results from both critical theory's inadequate conceptualization of power and its consequent misunderstanding of the relationship between power and emancipation. The more ample view of power developed through the works of Veblen and elucidated by Foucault provides a better understanding of the relationship between power and emancipation, increases critical theory's explanatory potential, and suggests that the right kinds of public bureaucracies may help to advance greater freedom to individuals than has yet prevailed.

Traditional Critical Theory and the Administrative State

As we discussed earlier, critical theory traditionally argues that the administrative state is inherently dysfunctional, oppressive, and conflict engendering. Administrative organizations are designed to procure ends and implement decisions that originate in society-wide processes of domination.[2] Domination is an exercise of power made possible through the colonization of individual lifeworlds by instrumental, scientific ideologies.[3] Discourses embodying instrumental reasoning are the mediums for exercising this power and securing this domination.[4] They accomplish this both by imposing specialized categories, technorational logics, means-end calculations and routinized techniques upon individual lifeworlds, and by ignoring or explaining away deviant features of life. When the attainment of assigned goals or the implementation of imposed decisions is frustrated, administrative action reduces to an exercise of coercive power, controlling, containing, or removing the source of resistance. Thus, by virtue of both the control they exercise and the purposive rational action they employ, administrative agencies limit what is really possible, constitute themselves a form of domination,[5] and establish loci of power struggles originating in and reinforcing society-wide processes of domination. In these ways, administrative agencies both reinforce the image of their expertise[6] and contradict individual and group interests in emancipation.[7]

Most important for our purposes, the technorational desideratum of consistency requires that a society administered by public organizations be erected upon the foundational assumption that citizens are utility maximizers functioning within a technorational consumerist system.[8] Successful navigation through such a system requires that individuals not only think technorationally and unambivalently in means-ends categories, but trust their fate to large-scale organizations that administer the apparatus that is rationally and scientifically constructed to pursue rational consumerist ends. Thus, bureaucratic power, legitimated through dominant discourses, subjugates individuals subtly but nevertheless aggressively, until they are denied both their true interests and their ability to think in any other way. In fact, the idea of individuals as free, reflective, creative, and self-determining agents is called into question as a construct engendered by the dominant technorational ideology and discourses embedded in public agencies.[9]

Together, these characteristics of public organizations restrict the exercise of free choice to smaller and smaller, more and more marginal spheres of public and private life. As the administrative state grows, administrative discretion increases. Administrators determine policy outcomes by controlling the flow of daily activities, specifying the details of programs, implementing agendas, and defining regulatory language. They direct our efforts through narrowly technorational definitions of what our problems really are; trigger routinized, legalized, and rationalized appropriate responses; and either impose sanctions or deny aid and reward should we fail to regulate our behavior accordingly. In this way, an administrative state where government is dominated by bureaucracies that have taken on a self-sustaining, self-directing, life of their own becomes inherently hostile to human freedom and self-determination. "The pattern of all administration tends of its own accord ... toward Fascism."[10]

Traditional Critical Theory, Power, and the Administrative State

As its view of the administrative state indicates, critical theory's understanding of power falls within a familiar conceptual tradition that may be thrown into sharp relief by a few of its salient works. These works constitute neither the extent of the debate nor the panoply of ideas over the meaning of power within the tradition. But they do provide a sufficient delineation of the tradition to justify the inclusion of critical theory's understanding within it and to distinguish Veblen's understanding from it.

In "The Concept of Power," Robert Dahl articulates the elemental notion of "power" in this tradition.[11] Simply put, power is the ability to get people to do something they otherwise wouldn't. Power is thus a characteristic of conflictual relationships in contradistinction to the contractarian tradition of

Hobbes, Locke, Parsons, and Arendt.[12] The latter tradition understands power as an ability or capacity that might be held, transferred, or utilized. Thus, individuals may hold power by themselves alone, or transfer power to others to act on their behalf, or generate power through cooperation and employ it to attain particular ends. Individuals utilizing their power may come into conflict, but the conflict is not inherent to the power itself.

According to the tradition of Dahl, however, individuals cannot hold power; it is only extant in exercises against another. That is, power obtains only when people would have acted differently and only when definitive action is taken in order to induce a desired change in their behavior. This tradition is also distinct from the "structural determinism" of thinkers like Poulantzas, who retain the relational view but condition the existence of power on a "system of material places occupied by particular agents."[13] Under this view, power is an attribute of total systems, individual behaviors being determined by the role of the person in the overall structure. For Dahl, no such inclusive system is necessary. "Nonenrolled" individuals may utilize power; all that is necessary is another with a contrary disposition.

Bachrach and Baratz initiated much of the debate over "power" within this relational tradition, arguing that the concept should include both getting others to do what they otherwise would not and the ability to prevent people from doing things they otherwise would.[14] Bachrach and Baratz include within the exercise of power the ability to create or reinforce such values and institutional practices as limit the scope of public debate and action to relatively innocuous issues. Although apparently at odds or at least somewhat divergent, both ideas nevertheless characterize power as a conflictual relationship among self-determining agents consciously advancing their individually defined interests against the understood interests of equally self-determining others.

Carrying on in this tradition, Lukes maintains that power is also exercised when people willingly do something against their interests. According to this view, power can operate "unseen."[15] That is, it can operate both through the control of situations and thus over perceptions, and through a control over what is possible and thus over behavioral dispositions. In these ways, the very desires and goals of people may be manipulated so as to facilitate their acting voluntarily in ways that are not self-defined and quite often contrary to their real interests, objectively understood.

Critical theory adds to this tradition the idea that power may be exercised to control not only the actions, desires, goals, perceptions, and behavioral dispositions of others, but paradigms and epistemologies as well. According to critical theory's version, this dynamic occurs not because of the intent of any group or individual, but because of the relationships of domination (structural, ideological, and discursive) that determine the routines of daily activity among the groups. Again, though the views of those in this tradition are somewhat at odds, all of them conceptualize power as both completely negative and operative through sets of social (interpersonal) relationships characterized by certain individuals and groups regularly securing their interests, goals, and desired outcomes at a noticeable cost to others.

In this critical extension of the tradition, domination may be discerned whenever the individual's desired outcomes and the means of attaining them are prescribed.[16] Thus, public administrative agencies are arguably the loci of domination as their goals, interests, preferences, and desired outcomes mediate the underlying power distribution ultimately founded upon society-wide structural, ideological, and discursive domination.[17] By transmitting sets of decision premises and cognitive expectations that constrain choices to those prepatterned by the dominant ideology and discourses, public organizations exercise a discipline over both perception and understanding that induces individuals to perpetuate relationships of advantage and disadvantage in their day-to-day "rational" decision making.[18] In Habermasian terms, public bureaucracies induce coordinated action outside of the "lifeworld," developing by themselves a structure of expectations, cognitions, and epistemologies (reflecting society-wide relationships of domination) beyond the interests of the individuals and groups to whom they administer.[19] Thus the exercise of domination through the power of public agencies inhibits the capacity of individuals to act as creative, reflective agents free of misconceptions about their own interests.

Traditonal Critical Theory, Emancipation, and the Administrative State

Despite the apparent totality of oppressive power operating through the relationships of domination in society, critical theorists currently insist that people can be emancipated. Emancipation requires both liberation from ideologies, power relationships, limiting paradigms, and constraining epistemologies, and an empowerment through a transformative fusion of theory and practice that is itself critically reflected upon both society-wide and in situ.[20] Thus, "emancipation" is a complex concept involving the absence of domination, an authority once denied, and the ability to apply certain capabilities (e.g., to employ available yet unfettering means to desired ends, to transform dominating practices and discourses, to evolve new emancipatory practices and discourses). In brief, emancipation involves releasing people to exercise power over their thought processes, lifeworlds, and anyone who would intrude upon them, while empowering them individually with the abilities necessary to realize their true interests in practice.

How to achieve this has always been a problem. Critical theory argues that people understand their true interests only upon emancipation from dominant ideologies, epistemologies, practices, and discourses. Immanent critique is the primary method employed to alert people to their domination. But domination, ipso facto, seems to neutralize immanent critique. Domination distorts consciousness, controls practice, and directs discourse, leaving people without the possibility of either discovering anything that is not already posited in the dominant ideology or understanding anything except through its prescribed epistemologies, practices, and discourses. In brief, all consciousness is a totalizing tautology.

Just as individuals lack recourse to the prescribed alternatives, groups and institutions are incapable of generating undistorted options. In fact, group participation and recourse to institutions transfers the prerogative of generating alternatives to some collectivity, subjecting individuals to another loci of domination and leaving them worse off than before. As dominated agents, then, neither individuals nor groups can enlist either public agencies or any other authority or expert in any project of emancipation that is not predefined by the society-wide practices and discourses of domination. By the same token, because public agencies, experts, and authorities arise from, mediate, and express practices and discourses of domination, they cannot themselves be so constructed as to aid in the emancipation of individuals and groups (at least not prior to emancipation). Thus, critical theory becomes understood as "the melancholy science"[21] and the Frankfurt theorists, as a consequence, must turn to pessimism, nihilism, and abstract negation.[22] The most that can be hoped for is that the Habermasian appraisal of the potential for emancipatory discourse and practice in the current dominant ideology will be realized in practice, ultimately reconstructing administrative and authoritative structure and practice toward emancipatory ends.

Problems and Paradoxes

For our purposes, the most important problem in traditional critical theory stems from the fact that it depicts the entire debate in organization theory over how to best achieve "responsive," or "democratic," or "reflective," or "educative" public agencies as misconceived. Given its conceptualization of power, domination, and emancipation, there seems a certain futility in attempting to invent or shape public institutions capable of enabling individuals and groups to live not just cooperatively but in a freely creative way with conflict, dissent, and the inability to reach consensus. Yet many of the freedoms individuals experience depend upon extant institutions. As indicated above, emancipation in critical theory includes not only the absence of interference but also the effective power to act. This suggests that forgoing a certain amount of self-rule may be necessary to organize effectively in order to accomplish things one cannot do alone. Therefore, both allowing interference with personal choices and allowing others to make certain choices for one may sometimes be necessary to be free.

For this reason, critical theory is in a quandary regarding public institutions. Marcuse, for example, argues that "all domination assumes the form of administration."[23] Power transmutes into "salaried members of bureaucracies who their subjects meet as members of another bureaucracy. The pain, frustration, impotence of the individual derive from a highly productive and efficiently functioning system in which he makes a better living than ever before."[24] So individuals are always dominated when presented with prepackaged options and they always suffer as a consequence. Yet, as Marcuse also notes, they live better than ever before. As this is the case, critical theorists sometimes recognize a need for "objectifying" a rationalized "lifeworld" in social and political institutions.[25] As Marcuse also says, freedom is a form of domination wherein the means provided (by bureaucracies) satisfy the needs of the individual with a minimum of displeasure and renunciation.[26] So Marcuse must recognize that, in some sense, freedom (at least from want) is institutionally dependent as both the needs of individuals and their capabilities (and hence the minimum necessary renunciation) vary with the level of economic and cultural development of those institutions. More definitively, Habermas ties the very possibility of emancipation to the emergence of more enabling political institutions made possible by the currently dominant ideology.[27]

Thus, while the forms of power acting through the administrative state are fundamentally at odds with emancipation, public agencies are among the instruments that can lead to the recognition and preservation of interests that must be addressed to secure emancipation. Sorting this out must involve discerning unnecessary constraints,[28] which will vary with the totality of objective and subjective conditions. But overall, to make everything consistent, traditional critical theory seems to require either the capture of administrative agencies as social (as opposed to political) institutions or a "minimalist administrative state" informed by and responsive to a radically democratic public sphere that operates according to the principles of the ideal speech situation. Still, all of this seems impossible given the negative nature of power and the domination it entails. Indeed, the idea of an "enabling public agency" seems oxymoronic. Ultimately, then, power and its relationship to emancipation as conceptualized by critical theory, entails an "enabling public agencies" paradox in practice.

A second problem arises from the fact that at the heart of critical theory is an assumption that reflecting upon a clear vision of extant power relationships, practices, and discourses is a sufficient emancipatory force. Reflective reasoning may be inhibited and distorted by administrative agencies, or by the pleasures and satisfactions of consumerism, or by religious ideologies, but ultimately the unhappiness, frustration, suffering, or alienation experienced as a consequence of being dominated and unfulfilled will spark the thinking required for emancipation.[29] However, recent critical scholarship suggests that change may require more than simply a clear head and the proper occasion for improved reasoning. The very ability to reason and to respond to reasoned argument may be inhibited by one's "cultural identity."[30] As "historical, embodied; traditional and embedded creatures," people may experience dominant practices and discourses as profoundly anchored "somatic" realities constituting the "deep structure" of the individual's identity and experience of "self." This, of course, significantly affects critical theory's hopes for emancipation. Both the liberation and the empowerment dimensions of the concept become questionable and another paradox is born. In practice, people suffering the most obviously unnecessary oppression and offered the most obvious institutional solutions often embrace and defend the oppressing ideologies and practices—hence the "conservatism" paradox so dispiriting to traditional critical theorists.[31]

A third problem arises from another of critical theory's foundational assumptions to the effect that domination is "arbitrarily" created by historical happenstance. Consequently, the relationships, discourses, and practices through which power operates may be done away with relatively easily given the proper perceptions, circumstances, and will. In this sense, critical theory anticipates transformation, not evolution. But it seems more likely that previous institutions, discourses, practices, and relationships both served some human interests at one time and provide an evolutionary foundation upon which reconstructions can be built. Consequently, as Wittgenstein demonstrated, though many discourses are possible, not just any will do. The terms of any "new discourse" must bear a reasonable "family resemblance" to previous usages of its terms if anyone is to understand it at all.[32] Likewise, social practices and institutions can be structured in many ways. However, just as the foundation of a building is connected to the structure erected upon it, so is the existing cultural base linked to social practices and institutions. Early discourses, practices, and relationships grew out of simple biological needs and of course took many forms, depending upon exigency and environmental situation. Nevertheless, current forms are built upon and still attached, however indirectly, to these foundations. To critical theorists, however, this is simply another instance of that paradoxical "conservatism" they find so dispiriting.[33]

A fourth problem arises from the assumption that the exercise of power through dominant practices, ideologies, and discourses as mediated by public agencies is capable of a totalizing impact. However, much scholarship indicates that the "bounded rationality," informal group processes, and psychological needs of bureaucratic actors prevent public agencies from functioning as efficient machines of domination.[34] Instead, they "muddle through" and perhaps function better when employing noncoercive organizational structures and decision-making processes.[35] Outside of the agency context, a number of interesting studies indicate that in practice the dominated (e.g., ethnic and religious minorities, homosexuals) manage enough power to establish and maintain their identities through complex sociopsychological coping mechanisms.[36] Moreover, individuals and groups may employ the dominant vocabulary and discourses in inventive ways.[37] Consequently, diverse, sometimes contradictory discourses may evolve and vie for acceptance within the dominant regime of meaning and practice, thereby enabling individuals to act accordingly. Abel and Marsh, for example, describe how dominated individuals and groups have historically adopted dominant vocabularies with subtle variations in the meanings of key terms in order to advance a variety of causes within American legal institutions.[38] Again, these instances of "interstitial emancipation" in practice can only seem paradoxical to traditional critical theory. They shouldn't be possible, yet they happen.

In sum, critical theory's negativity toward the administrative state derives ultimately from its conceptualization of power. This conceptualization accounts for its ambivalence toward the role of institutions in the emancipatory process, for its overreliance on reason as a sufficient emancipatory force, for its idea that extant relationships of domination are simply dispensable, and for its view that the power exercised through domination is totalizing in effect. In turn, these assumptions produce paradox upon paradox in practice.

Evolutionary Critical Theory and the Administrative State

We have demonstrated how Evolutionary Critical Theory closes the theory-practice gap while retaining critical theory's traditional critiques (e.g., of positivism and scientific reasoning), methods (e.g., deconstruction and immanent critique), and goal of emancipation.[39] Arguably, then, theory-practice gaps result primarily from critical theory's concept of power, and Evolutionary Critical Theory provides some concept of power that resolves the paradoxes of "interstitial" emancipation, "conservatism," and an "enabling public agency." It accomplishes these resolutions by positing a unified field theory of power. This theory of power retains the negative relational, conflictual dimension observed and analyzed by traditional critical theory;

reintroduces the personal, cooperative, and positive dimension observed by Hobbes, Locke, Parsons, and Arendt; and encompasses the "total structural" insights of Poulantzas. Furthermore, this conceptualization might not only resolve the paradoxes engendered by critical theory, but also might increase its explanatory power while retaining its critiques, methods, and ultimate goal.

Power, Emancipation, Veblen, and Foucault

Veblen examines the question of power first through a study of "the leisure class" and then through a study of "the vested interests" (a more inclusive term encompassing the leisure class). As a result of these studies, Veblen conceives of power as both much more diffuse and much more personal than does traditional critical theory. Power is found in individual "beingness," society-wide patterns of domination, and everywhere in between. Power is an ontological concept linking both different states of being (individual, social, institutional) and potentialities with actualities in the day-to-day flux of contending possibilities and tendencies. As a result, Veblen has very different ideas about where power is found, how it relates to subjects, how it is exercised, how it may be discerned, and how it is evaluated.

To begin grasping the nature of power, Veblen directs our attention to three everyday experiences. First, a person's mere existence elicits attention. Just the fact of our being engages a cluster of other people in reacting to our actuality, in taking us into account, and in compensating for us. This is an unintended capacity to produce effects that impacts those around us regardless of whether it is thoughtfully employed. While just our presence engenders reaction, we also experience unintended reactions to our natural propensity to act. "Every person is born a bundle of potentialities . . . [and delights] at the maturing of these potentialities into powers [as he or she begins to] talk, to crawl, to walk, to run."[40] As Veblen puts it, "as a matter of necessity . . . man is an agent . . . a center of unfolding, impulsive activity . . . seeking some . . . impersonal end."[41] This "instinct of workmanship" is a second unintended capacity to produce effects because it requires others to cope with us.[42]

As our existence inevitably embraces a stream of other people, we become known and know ourselves through the trail of these two simple experiences and their multiple effects through life. Of course, we react and adjust to the effects, to the "feedback" of reactions by those around us, but we also affirm ourselves. We define our boundaries synergistically with others, but we also put others on notice and engender an identity in the lives of others. Throughout his "Introductory" to *The Theory of the Leisure Class*, Veblen describes the many forms of unfolding taken by this personal power (this "instinct of workmanship") that everyone experiences—the power of being affirming being.[43]

May points out that in addition to the reactions others inevitably have to one's being, people subconsciously seek out opposition just in order to actualize this power.[44] That is, "being is manifested only in the process of actualizing its power . . . power becomes actualized only in those situations in which opposition is overcome."[45] Veblen also notes this aspect of "being power." It is the innocent origin of "invidious discriminations" as esteem is gained by putting one's efficacy in evidence.[46] Thus, Veblen's "personal power" has a dimension akin to Nietzsche's notion that "to impose upon becoming the character of being—that is the supreme will to power."[47] It anticipates Tillich's ontological claim that "every being affirms its own being . . . even if its self-affirmation has the form of self surrender."[48] Moreover, it is echoed in the psychoanalytically identified archetype of striving for the solution to life's problems as encountered in the evolution of the individual.[49] In brief, Veblen's "personal power" is the capacity people have to produce unintended effects around themselves simply by virtue of their being; and opposition (resistance) followed by the continued force of being, the continued self-affirmation of simple persistence, is a natural (unintended) expression of this capacity.

Finally, Veblen calls attention to the third shared experience of reality imposing itself upon us. Against this we act out both physically and psychologically. That is, personal "being power" may be employed toward the gratification of our interests in opposition to the frustrations of reality. So in addition to unintended effects, "being power" may have intentional effects. Veblen understands clearly that these "intentions" may be manipulated and dominated by extant ideologies, discourses, and practices. These are elements of the complex reality imposing itself upon us. Nevertheless, it is each of us as an individual who is asserting these interests, and that is, in and of itself, an actualization of our personal "being power."

As we have seen, critical theory understands power as a society-wide phenomenon not attributable to any individual. However, Veblen's observations concerning personal "being power" justify his closing of the theory-practice gap by focusing on the "microphysics of power." That is, they indicate that to really understand the nature of domination, to discern its actual origin and to accurately depict how it works, critical theory must expand its scope to include a critical analysis of the everyday, one-on-one relationships and struggles of simple interpersonal being. These observations further suggest that they should not be approached as echoes of an overarching total domination but as the very "stuff" of domination and emancipation that is built up into society-wide practices through emulation, repetition, and habituation.

Fundamental to Veblen's analysis is an understanding of the diffusion and reactivity of power as explicated by Foucault. Like Veblen, Foucault understands power as neither a "group of institutions and mechanisms . . . [nor] a mode of subjugation . . . nor a general system of domination."[50] These are only certain elaborate forms that power may take. Power originates in any relationship where force is employed.[51] That is, if we imagine a domination-free context, as each person meets another the necessity of adjustment arises; and this entails necessarily doing something we otherwise would not do (even if it involves getting other people to do something they otherwise would not do or keeping them from doing what they otherwise would). Even in contexts of domination, when socially prescribed meetings occur (e.g., guard and prisoner, doctor and patient, professor and student), the tactics of employing power in situ produce shifts resonating in the overall pattern of power. They produce a sort of unintended and often unnoticed "drift," as each partner to the relationship compensates for the context and copes with the other under the exigencies of circumstance. As Foucault puts it, "power's condition of possibility . . . is the moving substrate of force relationships which, by virtue of the inequality, constantly engender states of power, but the latter are always local and unstable."[52]

Veblen identifies four sources of drift.[53] First, because people are active agents by virtue of their being power, they manipulate, adapt, and mold dominant power relationships, practices, and discourses to achieve various objectives (many of them culturally dictated) in situ. As a consequence, unusual interpersonal behaviors, with uncertain consequences, will normally arise without comment. If emulated and summed across all individuals, larger shifts in practice, discourse, and power relationships emerge. Next, the sheer complexity resulting from both such cumulative emulation and the simple diversity of where individuals and groups find themselves in the social structure at any given time always yields some novel reaction to the dominant ideology, its practices and discourses. Third, attempts (successful and unsuccessful) by institutions mediating the dominant ideologies and practices to normalize the impact of both the "instinct to workmanship" (the natural "acting out" of human beings) and the unique patterns of interaction necessary for solving unforeseen problems produce their own "drift." That is, the attempt to keep "drift" from happening, partially by realistic considerations, partly by the mythical beliefs of "imbecile institutions," itself constitutes a drift, sometimes by misdirecting the intentional behaviors of individuals. Finally, "backward drift" may occur.[53] Where the dominant practices and discourses attempt to cope with information or experiences unaccounted for in the dominant ideology by reinterpreting, ignoring, or suppressing them, the blind lead the blind. The resulting ignorance of what is really happening results in a "drift" to waste and ineptitude.

Including this dimension of power resolves the paradox of "interstitial emancipation." Veblen and Foucault's unified field conceptualization of power disputes the totalizing impact of the dominant ideology on both individual thought and the circumstances under which people act. Throughout *The Theory of the Leisure Class*, Veblen is at pains to demonstrate how the prescribed practices of the dominant capitalist ideology instill proclivities in people for both activities and commodities that enlist their enthusiastic participation in maintaining their own domination. Thus, for example, the practice of invidious discrimination enjoins the conspicuous consumption of peculiar clothing, lawns, delicate women, and pedigreed, largely dysfunctional dogs. However, these are proclivities only. While people may be "encouraged" tacitly and subtly to think and act in certain ways,[54] their personal "being power" and the exigencies of circumstance constitute a moving substratum of power productive of "drift."

But this personal "being power" is not all there is to power. From these unstable, drifting, necessarily heterogeneous, one-on-one force relationships are thrown up patterns of interaction that, if repeated often enough by enough people (i.e., if "strengthened" through "emulation"), become habits (Veblen) and strategies (Foucault) eventually acquiring the appearance of a coordinated whole emanating form a single source.[55] Crystallizing habits and strategies eventually institutionalize into state agencies and "social hegemonies"[56] that may then be mobilized against the drift. That is, widely practiced "habits of thought" (Veblen), patterns of power (Foucault), and practices of domination may be fed back, conditioning both "being power" and one-on-one relationships of force, inhibiting change, and stabilizing certain emerging patterns of power at the expense of others.

At the same time, even as institutionalized "social hegemonies" inhibit change and stabilize patterns of domination, they reveal "truths" and thus may empower and emancipate. "Truth is not the reward of free spirits . . . [it is] a thing of this world . . . produced only by virtue of multiple constraints."[57] We do not know how imprisonment works, for example, or its effects on people until we imprison. We do not know the nature or extent of the emancipation or domination an institutional structure may provide until we establish it. Hence, the truths we know depend upon the patterns of practices constituting our society's "regime of truth,"[58] and what we are empowered to accomplish is similarly dependent.

Veblen expresses this same complex of ideas in his theory of endogenous evolution. Social evolution, according to Veblen, does not proceed through a selection of traits by the external environment. There are no "exogenous constraints" or selection mechanisms outside of the social system. Change emerges from internal variation through endogenous forces, "drift," cumulative emulation, habituation, and institutionalization. External influences do affect the system but they must be included in the system to matter to the people in it, and the understanding of those external influences is mediated by the dominant ideology, practices, and discourses of the system.[59] For this reason, evolution is not necessarily progressive or emancipating. Societies may fail to "internalize" exogenous forces; they may misconstrue them or otherwise fail to replace obsolete understandings. Endogenous "imbecile institutions" might inhibit or preclude effective response to internal or external forces, and endogenous changes might themselves produce patterns of response and emulation that the system cannot respond to in a progressive, emancipatory fashion.[60]

This has two important implications. First, because what we can know and accomplish is limited by our "regime of truth," sooner or later we will run up against the limits of that regime. We will come across something, or feel something, or something will happen that the regime of truth handles at best very clumsily if at all. We will encounter something the regime enjoins to silence or affirms as nonexistent.[61] The things that are not spoken about become obvious by their omission from discourse. Veblen certainly understood Darwinism as just such an example of how, sooner or later, one runs up against the limits of established accounts. While contradictions between experience and ideology (the fossil record versus the biblical dating of creation), and such incongruities as dinosaurs, may be accommodated or explained away by dominant epistemologies and paradigms (Genesis 6:4: "there were giants on the earth in those days"), some slight redefinition of reality, some drift, occurs. The mere fact that evolution is asserted requires an accommodation, even if it is only to ignore it or enjoin people to silence about it; and denying change is itself a change.

The second important implication is that because "we cannot exercise power except through the production of truth, . . . power never ceases . . . its acquisition of truth: it institutionalizes and rewards its pursuit."[62] Under this conceptualization, power can emancipate because emancipation does not imply the absence of constraint, as in the conceptualizations suggested by the contractarian, structuralist, and critical theory traditions. Rather, emancipation requires the presence of certain abilities and supporting conditions, a certain "regime of truth," that makes agency possible. On the one hand, then, Veblen and Foucault's concept of emancipation is true to the critical theory tradition because it encompasses the empowerment to realize individual interests in practice. Contrary to traditional critical theory, however, it does not see power as the enemy of emancipation and has no ambivalence about the positive role public agencies can play in emancipating. They are instituted in pursuit of truth, dominating and emancipating simultaneously.

Clearly, including this understanding of how power works resolves the oxymoronic paradox of "enabling public agencies." But it resolves the paradox of "conservatism" as well. Persistent conservatism in both American and European societies, despite the advance of democracy, was a significant factor in the Frankfurt school's ultimate loss of hope for a final emancipation in practice. The thinkers of that school attributed conservatism to the power of both the growing administrative state and the media-based domination of a capitalist ideology that produced a mass consciousness suitable for the market economy and amenable to control through the promise of consumer satisfaction.[63] In the face of what they took to be the totalizing impact of these two factors, the Frankfurt theorists were incapable of suggesting a critical praxis other than Marcuse's[64] endorsement of violence as a "natural right" of oppressed minorities.[65]

Veblen, however, understands conservatism as a matter of necessity. It is a matter of building upon the regime of truth already in place through a cumulative process of "internalizing" exogenous and endogenous forces "that cumulatively change as the process goes on, both the agent and the environment being at any point the outcome of the last process."[66] It is also the result of the complexity of forces "overdetermining" individual choices, the different forces acting upon people occupying different places in the social structure and the resultant acts of force rippling through society. Leisured individuals, for example, are loath to change because their comfortable position insulates them from perceiving the need. Moreover, others in society accord honor to their habits and grant them authority. All of this joined with a general hesitancy regarding the unknown and the emotional stress and physical exertion necessarily involved in any (even attractive) change results in an understandable inertia. Hence, complex economic, social, psychological, and political forces that no one intends encourage individual decisions perplexing behavior in ways that reinforce patterns of dominance. The inertia of the leisured in turn becomes a force, obvious to everyone and requiring accommodation in the course of everyday interaction. The force of this inertia thus ripples through the "lower classes," who not only fear the unknown and eschew the emotional and physical exertion necessary for change, but lack innovative energy because it is sapped by the necessity of working long and hard to make ends meet. Those in the middle classes who have "surplus energy" expend it in the pursuit of "conspicuous decency," a response to the social forces encouraging people to meet certain standards of reputability and social acceptance. Finally, everyone's behavior feeds back to everyone else, who interpret the behaviors of the others as a vindication of the status quo, either as just the way things are naturally or as the way they must be in order to secure a well-ordered society.[67] Everyone thus responds to a complex of forces emanating from everywhere at once, and everyone acts intentionally in ways that unintentionally reinforce the distribution of advantage and disadvantage. "Drift" occasioned by "being power" and the adjustments made in the moving substratum of force relationships are the forces of change. They, too, are intentional at the microlevel but unintended in their systemic effects.

Conclusion

Critical theory has foundered until now because of its inadequate understanding of power and how it works. Understanding power as a society-wide phenomenon not attributable to any individual, as totally negative, and as totalizing in impact as exercised through the institutions of capitalism leads to both an assumption that the administrative state could not but dominate and an abundance of paradox. However, resolving these paradoxes using the perspective of Evolutionary Critical Theory allows us to revisit, reexamine, and reconsider the possibilities that exist. When we take this different view of power, which allows us to see both the positive and negative possibilities, we can then begin to understand how the potential for a positive conception of an administrative state can exist and can function as a means to craft a "good society."

Evolutionary Critical Theory resolves the paradoxes associated with the more "traditional" views of power and both establishes and galvanizes the emancipatory role for public agencies by conceiving of power as an ontological concept both linking different states of being (individual, social, institutional) and connecting potentialities with actualities in the day-to-day flux of contending possibilities and tendencies. Evolutionary Critical Theory perceives power as coming from everywhere, circulating in all directions, unifying everyone affected into a single social field (in essence, framing our intersubjective experiences), constructing individuals and groups, revealing those constructions to us, and establishing the conditions and contexts enabling people to act as agents or inhibiting their ability to do so. Every institutional situation and public agency is similarly a product of power circulating all about from everything that precedes and is concurrent with it, and may similarly help to advance or retard emancipation.

The resulting totality is a sort of organism within which everything is explained by the power relationships and practices of everything (and everybody) else. People, groups, institutions, and all other relational forms are both agents and "dependent variables." Emancipatory effects, regression, inertia, unification, and the totalities of structure and place emerge from the intentional acts of agents, while the overall pattern of power relationships and the resulting institutional and ideological domination and opportunities for emancipation are unintended. As people are both sources and symptoms of power, power is both more personal and more diffuse than understood by traditional critical theorists. Incorporating these evolutionary dimensions of power into critical analysis greatly enhances its explanatory power and uncovers potentials for emancipation that traditional critical theory overlooks.

As a consequence, we can then begin thinking about how an individualistically focused, radically democratic, minimalist society devoid of power relationships might lead us toward the goal of a "good society," as many critical theorists might contend. Throughout this treatise, we have increased our understanding of how eliminating teleology, understanding power, and having a discipline with ontological status allows public administration to express the synergy among several social forms including individuals, institutions, and experiences. This synergy can then lead to emancipatory outcomes, forcing us to consider how we might effectively achieve the goal of a "good society," one of the highest goals of critical theory within institutions and power relationships.

7

Evolutionary Critical Theory and the "Good Society"

We have established earlier that public administration often struggles with legitimacy issues based on how theory is crafted, how its ontology is understood, and how power is wielded both conceptually and practically in the discipline. We have addressed both sides of the arguments for each of these issues earlier in this book. The ultimate consequence of resolving these issues places public administration, both as discipline and practice, as the nave for moderating the intersubjective experiences in any conception of a "good society." Critical theorists have long argued that bureaucratic structures mediate dominant ideologies, practices, and discourses, thus keeping us from realizing a good society by frustrating emancipation and perpetuating patterns of advantage and disadvantage. We argue that such an outcome is only one of a number of possible scenarios. Furthermore, this mainstream position proffered by some theorists is founded upon an untenably individualist notion of emancipation that frustrates any possibility of movement toward a good society, even as understood by critical theory. We argue that Evolutionary Critical Theory enhances our understanding of the good society and offers a more tenable understanding of what it means to be emancipated. It thereby suggests a legitimate role for bureaucracy in a good (critical) society. To accomplish this end, we must understand public administration's roles, the nature of a good society, the limits of traditional critical theory, and how an evolutionary approach solves many of the "wicked" problems facing theorists in praxis.

A number of mainstream or orthodox theorists in public administration are predominantly motivated by a belief that bureaucratic organizations contribute to society by bringing order to an otherwise chaotic, confusing, and crushingly complex reality.[1] In this regard, they tend to follow Weber in his belief that "the purely bureaucratic form of administration is . . . superior to any other form in precision, in stability, in the stringency of its discipline, and in its reliability."[2] Veblen supports and extends these arguments by articulating how organizations, both public and private, have a usufruct within and among their organizational structures that bind and support the development of a viable community.[3] Given the multiplicity of such views, bureaucratic institutional forms and practices are necessarily legitimate and necessarily integral to any good society, forcing us to consider if emancipation can happen within such forms and practices.

Contrarily, many critical theorists seek a good society through neutralizing bureaucratic institutional forms and practices as much as possible. Their belief, in Habermasian terms, is that public bureaucracies induce coordinated action outside of the lifeworld, developing by themselves a structure of expectations, cognitions, and epistemologies (reflecting society-wide relationships of domination) that are incompatible with the interests of the individuals and groups to whom they administer.[4] Thus, the exercise of domination through the power of public agencies inhibits the capacity of individuals to act as creative, reflective agents free of misconceptions about their own interests. As both bureaucratic forms and practices are thus among the cultural impediments to emancipation, both must be either eliminated or radically circumscribed in order for a good society to emerge.

While critical theorists recognize the danger of domination inherent in bureaucracies and bureaucratic practices, they are somewhat ambivalent about their ultimate legitimacy. While Marcuse, for example, argues that "all domination assumes the form of administration" and that "power transmutes into salaried members of bureaucracies who their subjects meet as members of another bureaucracy," he also notes that because of such bureaucracies people live much better than ever before.[5] As public bureaucracies thus serve some real interests by emancipating people from want to an important extent, critical theorists recognize the value in sometimes "objectifying" a rationalized "lifeworld" in social and political institutions.[6] More definitively, Habermas ties the very possibility of emancipation to the emergence of more enabling political institutions made possible by the currently dominant ideology,[7] and Honneth argues that it is the granting of respect within social and political institutions that lays the groundwork for individual emancipation.[8]

We argue that this ambiguity among critical theorists about whether a good society might be realized in a context of bureaucratic forms and practices arises from an untenably individualist notion of emancipation and certain consequent misunderstandings concerning (1) the relationships among individuals, (2) the relationship between individuals, and (3) how both types of relationships work toward a good society. Furthermore, the less individualist view of emancipation held by Veblen and elucidated by Foucault provides a better understanding of the relationship between emancipation and the good society, explaining how a good society may be secured through power relationships and their mediating institutions. This, in turn, legitimates bureaucratic institutions as critical factors in any viable conception of a good society.

The Good Society and Its Dilemma

One important and popular purpose of political and administrative theory is to articulate comprehensive alternatives to the ways things are with a view toward establishing something everyone might recognize as a good society.[9] Now, any inquiry into the nature of such a society involves necessarily some consideration of how far we are comfortable with allowing individuals to decide certain things for themselves. One of the greatest debates in this regard is over how far we as a society are comfortable with allowing individuals to decide what is in their best interest and so what values they will hold and what particular lifestyle they will pursue. Broadly speaking, for example, a good society for those on the left side of the ideological spectrum is one that countenances state intrusion into such choices only to the extent necessary to ensure that individuals act as good citizens. Those on the right side of the spectrum tend to maintain that the state ought to intervene in the individual's personal life and value choices to the extent necessary to develop good people with the values, attitudes, beliefs, and interests necessary to making the society a good one.

Critical theory "is founded on the vision of a better world, and, simultaneously, the refusal to describe this utopian vision in positive, substantive terms."[10] At best, the structure and processes of its utopian society are hinted at through an ongoing critique of modern society.[11] Nevertheless, it is abundantly clear that, save for anarchist theory, critical theory is perhaps the most thoroughgoing in what it believes a good society must leave to individual choice.

Now, this respect for individual life choices entails a dilemma. Held as a primary value, it requires a good society to constrain both coercive and noncoercive governmental attempts to influence an individual's choice of the good life. However, a political order that does not promote some view of the human (common) good must deal with what Plato called an "emporium of constitutions" or ways of life, some of which will oppose individual choice in matters of lifestyle, values, attitudes, and beliefs.[12] In fact, no one legitimate view of the good society or the good life can be discovered empirically in any society where private decisions on such issues are given some play.[13] This circumstance, in turn, threatens to frustrate the realization of a good society by engendering conflict as individuals and groups strive to further their personally understood interests and personally valued lifestyles through the power of government and its bureaucracies.

In response to this dilemma, some theorists argue that a good society distinguishes between political virtues (the right) and personal virtues (the good), enforcing legitimately only those principles of "fair social cooperation" that citizens can agree upon as conducive to securing the requisites for realizing their personal values and lifestyles.[14] Others argue that a common understanding of the good society can develop without coercion when individuals are drawn out of the private realm by participating in public (political) life.[15] In both cases, the idea of a good society involves the development of a thoroughgoing and highly inclusive participatory democracy as a check on bureaucratic power and its extension into areas of personal choice. However, such approaches prove unsatisfactory. First, they tend to result not in a shared idea of what constitutes a good society but in little more than the "combining of preferences, all of which are counted equally."[16] Moreover, any claim that "a single legitimate set of values, attitudes, and beliefs will result from participation in public life implies that any society currently valuing personal choice in such matters ought to be replaced by the rule of some authority embodying the revealed truth."[17]

Another approach to resolving the dilemma is suggested by those who insist that "we cannot conceive our personhood without reference to our role as participants in a common life."[18] In this view, individuals do not choose their values, attitudes, beliefs, and lifestyles as unencumbered selves but as selves constituted at least in part by shared conceptions of the good.[19] As there is no unencumbered self, there are no purely individual values and interests and there is always some irreducible minimum set of interests and values that are shared. Under this approach, the pursuit of this minimum might be pursued legitimately through public institutions. Nevertheless, such thinkers tend to reject the idea that the self is "radically situated" and so incapable of any real choice, however encumbered it may be.[20] Thus, the dilemma is not really resolved.

Critical Theory's Resolution: Emancipation and Praxis

Critical theory's answer to this dilemma is praxis. The original members of the Frankfurt school, for example, sought emancipation from societal incursions into the individual's lifeworld through a social science that reflected upon the ordinary experiences of individuals as they struggle under bureaucratic and other power relationships mediating and reinforcing the dominant abstract ideologies, epistemologies, and practices of modern capitalist society. In this way, the Frankfurt school hoped to sensitize people to the "gaps" between what the dominant theory promised and what they actually experienced in practice on a daily basis. Awareness of these gaps, it was thought, would bring about a realization that by changing power relationships and practices, new understandings (new theories) of what a good society might be would arise and close the theory-practice gaps.[21] Habermas also sought emancipation by examining how the theoretical ideal was distorted in everyday interpersonal practice by dominant interests.[22] Focusing on communicative practice, Habermas described how the "ideal speech situation" was conducive to both truly effective praxis and the emergence of a good society and how it was systematically distorted by dominant interests to obscure extant theory-practice gaps. Similarly, Honneth and Farrell sought emancipation by looking to "those circumstances experienced as unjust . . . on the basis of the criteria that affected subjects themselves use to distinguish between a moral misdeed and mere ill luck."[23] Experiences of injustice or disrespect engendered by domination within relationships that were (1) close and personal, (2) political, and (3) civil or social were for Honneth the source of both any struggle for change and the normative standards of a good society to be employed in praxis.[24]

From critical theory's point of view, then, the apparent dilemma or contradiction between personal choice and societal integrity arises from the fact that most political and administrative theories are too concerned with articulating the intrinsic desirability of some proffered ideal.[25] Eschewing pragmatic appraisals of political hazard and insufficiently attentive to individual "lifeworlds" (especially those of minorities and the otherwise marginalized), dominant groups give rise to the dilemma as they attempt to implement an abstract ideal in the real world. The solution for critical theorists is to test theory through action, immanent critique, and other forms of deconstruction so that the dominated may realize the nature and source of their oppression, emancipate themselves from it, and decide independently what role (if any) is legitimate for public institutions in a good society.[26] Thus, although critical theorists are fearful of becoming themselves repressive purveyors of utopian blueprints for a good society,[27] it is clear that because praxis is the key to constructing and deconstructing society, and because praxis requires emancipated individuals, any good society must involve mechanisms for ensuring that individuals become and remain emancipated.

The Nature of Emancipation

The emancipation sought by critical theorists is a condition of uncontradicted individuality and unencumbered agency limited only by the need for reflective self-critique. It is through this "freedom of the detached self" that the individual "finds the meaning and ground of its existence in itself."[28] Thus, the emancipated individual pursues his or her true interests either regardless of dominant ideologies, practices, discourses, and bureaucratic institutions or in a manner highly skeptical of them. If familial, political, social, or economic conditions seem oppressive or disagreeably binding, people are invited by critical theorists to emancipate themselves from constraint, and to restore themselves to a state of freedom and independence.

Initially, for example, Horkheimer envisioned a society that "would neither hinder nor obstruct the complete development of individual human nature."[29] The result would be emancipated individuals capable of constructing, deconstructing, and reconstructing their own good societies. True to his critical ideals, Horkheimer went no further in describing his vision of the good society because "everything . . . depends on creating the free subject that consciously shapes social life."[30] Later, despairing in the face of what he took to be an inevitably triumphant modernity, Horkheimer urged a resurrection of religious faith as a way of emancipating the individual from dominating institutions and practices. Religion thus became Horkheimer's means to an "interior independence of this world" and a tool for the individual realization of personal emancipation.[31]

Similarly, Adorno's ultimate vision was of "a society in which the people have become autonomous individuals [freely] choosing their government and determining their life."[32] While he sought a "middle path" between the extremes of relativism and absolutism, to reconcile "unity" and "difference," his "utopia would be above identity and above contradiction."[33] Thus it required an emancipation from all forces of society that work to define, categorize, prioritize, direct, limit, or "contradict" in any other way the self-determination of the individual as self-critically self-defined. Marcuse's dedication to such an individualist emancipation was so strong that he was willing to accept it as an authoritarian standard,[34] and Walter Benjamin's search for an emancipatory aesthetic was engendered by his conviction that "humanity's negativity fostered its greatest achievements,"[35] that art speaks to the individual and that the realization of truth it brings is corrupted through communication.[36]

In this same vein, Habermas, in both his early and later work, suggests that whenever individual interests become community interests they threaten emancipation through their domination of discourse and practice. This, in turn, negatively affects the ability of individuals to critique society further and threatens to fossilize ideas of what constitutes the "good." Consequently, such constraints have to be removed, and Habermas offers first the "ideal speech situation" as a means to this end and then the courts as a corrective to the possibility that even in ideal speech situations marginalized and minority voices may be incapable of a meaningful influence by virtue of their nature, situation, and "lifeworld."[37] Honneth and Farrell evince the same emancipatory intent. Their goal is to work out the conditions necessary to realize Hegel's "sphere of individual freedom," wherein individuals are personally empowered with not only the legal rights and the social esteem necessary to realize their interests, but an "emotional security in the articulation of the claims raised by their drives" as well.[38]

Critical theorists, in brief, conceptualize emancipation as a state of disassociation from power relationships, dominating institutions and practices, and even limitations imposed by radically inclusive democratic processes. It is a state of thorough individualism wherein individuals pursue individual ends as informed by individual choice, individual behavior, and individual self-expression and limited only by individual reflective critique. Additionally, emancipation includes the empowerment to satisfy interests through a transformative fusion of theory and practice in individual lives that is itself critically reflected upon individually, in groups, and society-wide.[39] Taken altogether, then, emancipation involves releasing individuals to act as agents by exercising power over their thought processes, "lifeworlds," and anyone who would intrude upon them, while empowering them individually with the abilities necessary to realize their true individual interests in practice.

Problem: The Dilemma Returns

While praxis is critical theory's answer to the dilemma engendered by the fact that respecting individual lifestyle choice threatens to frustrate the realization of a good society, its concept of emancipation leads to the exclusion of positive critique and limits its practice to stressing ideological, class, race, gender, economic, legal, and social differences. As Benjamin warns, adhering to such a position (what he terms "the melancholy left") leaves one standing "not left of this or that orientation, but simply left of everything that is possible."[40] This position tends to promote the very fragmentation and discord that emancipation and praxis were meant to overcome.[41]

Consider, for example, that given critical theory's individualist understanding of emancipation, any good society must develop and maintain autonomous, initially self-concerned individuals and at most a "minimalist state," informed by and responsive to a radically democratic public sphere that not only operates according to the principles of the ideal speech situation,[42] but also grants every individual meaningful (efficacious) recognition.[43] To be legitimate, the state must be "minimal," as bureaucratic power must be kept from subjugating individuals to such an extent that they are denied both their true interests and their ability to think for themselves.[44] Thus, critical theory's individualist orientation encourages each of us to seek a good society by advancing personal interests for recognition and redress in a public sphere and admonishes each of us to grant everyone else an equal recognition. Maximally responsive and "minimal" social and political institutions then "objectify" a rationalized "lifeworld," ostensibly by transforming personal interests into policy outcomes. The resulting array of outcomes becomes a "good society."[45]

Now, the resulting society is necessarily "contingent." It is more or less satisfactory (legitimate) for the present to the particular individual interests expressed and considered genuinely in the public sphere at a given time, and it is always subject to change with individual "lifeworld" circumstances.[46] This follows because critical theorists anticipate that any set of genuinely expressed and considered interests and the resulting policies, institutions, and power relationships will be continually deconstructed through negative critique. In this way, society remains responsive to any change (internal and external) that negates emancipation. For this same reason, critical theorists do not seriously contemplate the possibility of a positive critique of extant power relationships and institutions. Positive critique is impossible as it necessarily applauds (legitimates) some set of power relationships and accordingly risks both an end of critique and the domination of some individual interest now or in the future.[47] Consequently, critical theorists are not only unwilling but also unable to suggest any direction we might take toward a good society.

The problem is that this indeterminacy minimizes critical theory's impact on practice, encourages conflict, and leads to individual and group marginalization, the very opposite of the good society that critical theorists seek. For example, while critical theorists from the original Frankfurt school through Habermas and Honneth hold that both identity and agency are formed intersubjectively, they seek a society wherein individuals emancipated from dominating intersubjective forces interact to form and reform themselves. How to achieve such emancipated intersubjective beings, given the social construction of identity and agency, has always been a problem. Fay, for example, points out that people are "historical, embodied, traditional and embedded creatures," experiencing dominant social and institutional practices and discourses as profoundly anchored "somatic" realities constituting the "deep structure" of the individual's identity and experience of "self."[48] Critiquing this deep structure is therefore likely to engender alienation from both society and the self and hence any source of agency. Similarly, Armour argues from his studies that personal identity is constructed through identification with both a social class and a "homeland," regardless of how deficient or oppressive the political, economic, and social institutions, practices, and discourses may be.[49] Thus identity, the sine qua non of agency, seems to depend upon extant practices, discourses, and institutions. Consequently, individuals tend to avoid disassociating (emancipating) from them as that may effect a loss of both identity and agency (in praxis, a reduction in practical or actual rather than conceptual freedom).

A similar dynamic occurs at the group level. As identity and agency depend upon group practices and discourses, it becomes difficult for existing groups to "internalize" critiques and changes that cannot be easily managed within current practices and discourses.[50] Consequently, those scholars critiquing and attempting to introduce or support the change are often marginalized and deprecated as defensive maneuvers. Even if efforts at change are successful and the group is emancipated from traditional practices and discourses, the new practices and discourses tend to marginalize those whose status, position, or acquired skills are devalued by the change.[51] In either event, attempted emancipation encourages fragmentation, marginalization, and opprobrium.

This marginalization may be intensified in the case of already marginalized groups even though they seek not a society-wide emancipation but simply a loosening of the norms as applied to them. In *Reynolds v. United States*, for example, the Mormon community sought emancipation from the dominant practice of monogamy. Mormons were already marginalized because of their religious and economic practices (e.g., communal sharing of goods and services), which forced them to relocate to remote geographic locations. Their persistence in the pursuit of liberation from monogamy pursuant to the First Amendment guarantee of religious freedom resulted in their increased marginalization and deprecation. Consider, for example, this brief excerpt from the case:

> Polygamy has always been odious among the Northern and Western Nations of Europe and, until the establishment of the Mormon Church, was almost exclusively a feature of the life of Asiatic and African people. At common law, the second marriage was always void, and from the earliest history of England polygamy has been treated as an offense against society.[52]

Clearly, this excerpt portrays Mormons, Asians, and Africans as odious groups whose practices must be marginalized, if not eradicated. The remainder of the decision makes it clear that their practices "threaten" the dominant culture and unless outlawed would destroy civilization as we know it.

Even when individuals and groups do manage to emancipate themselves from the practices and discourses of the society at large, the results are not unqualifiedly desirable. A number of interesting studies, for example, indicate that some communities with values, attitudes, and beliefs contrary to the dominant ideology (e.g., ethnic and religious minorities, homosexuals) do manage to establish and maintain their practices separate and apart from the society around them.[53] However, while this enables the individuals composing such communities to act independently, they remain marginal relative to society as a whole and never fully participate in the evolving social, economic, and political life of that society. Their agency is thus severely circumscribed.

In *Yoder v. Wisconsin,* for example, a Quaker sought the right of exemption from public school attendance for his children once they finished the eighth grade. The exemption was granted pursuant to the First Amendment guarantee of religious freedom, although the tenets of Quakerism do not technically require believers to keep their children from the schools (it was simply the community's practice). Thus, the children and the Quakers generally were emancipated from the practices and discourses of the dominant ideology. As the dissent in that case pointed out, however, the children were thereby denied access to both the common opportunities accorded everyone else and a significant measure of self-determination in the future.[54]

Equally important, however, is parental loss of exposure to the evolving society about them and the opportunities for agency that it offers. This phenomenon has been well studied in another context by Barnes.[55] Barnes found that so-called status groups regularly engage a number of devices to restrict their contact with outsiders.[56] He also found that this actually places them at a disadvantage as new technologies, ideas, or cultural advances lead to change. This phenomenon was a favorite topic in Veblen's *Theory of the Leisure Class,* in which Veblen discusses at length the sheltered position of the leisured, which both "emancipates" them from the rest of society and its concerns and renders them quite slow to adapt.[57] In the case of the leisure class, of course, its dominant position and relationship to the "pecuniary functions" kept it sufficiently in touch. In most other cases, however, while diverse, sometimes contradictory discourses and practices may vie for acceptance and receive at least tolerance within the dominant regime of meaning and practice, such success does not liberate them so much as reinforce their marginality and bind them to certain other practices and discourses.

In sum, to the extent that values, practices, beliefs, and discourses contrary to those that are dominant secure a place apart, they are, for better or worse, marginal to the "form of life" evolving about them. They thereby exacerbate social fragmentation. They are not embedded in everyone's daily discourse and activity. Our relationships and patterns of thought are not conditioned by them, they are not regularly involved with us, and the meanings of their terms and their arguments are neither as easily understandable nor as convincing to us once understood. Hence, they are outside our community of interest and play little part in lending meaning to our understanding of a good society. Of course, this could change. Minority communities of interest are endogenous forces whose practices and discourses may be internalized and employed by us, becoming part of our idea of a good society. Nevertheless, this will be by our agency, not theirs. It is the power and agency arising from active participation in the institutions and practices of the dominant "form of life" that makes all the difference.

What this suggests for critical theory, of course, is that in practice our capacity to act as agents (to be emancipated) flows from the very non-individualist power relationships traditional critical theorists seek to emancipate us from. This paradox was recognized by Horkheimer, who realized that critical theory's emancipatory techniques were effective only within the conceptual possibilities of the Enlightenment's power relationships and overall "form of life." As that "form of life" ended with the victory of instrumental rationality over the tradition of objective reason, Horkheimer concluded that immanent critique (and perhaps critical theory itself) had lost its conceptual context and was rendered ineffectual.[58] In other words, Horkheimer came to realize that detachment from (Enlightenment) social, political, and economic practices and institutions (in his case due to the end of a "form of life") renders critique and movement toward a good society impossible, because critique through and within praxis must end with that disengagement as well. Thus, no emancipatory practice seeking the freedom of the detached self, wherein the individual finds the meaning and ground of its existence in itself through a completely uncontradicted individuality and unencumbered agency limited only by the need for reflective self-critique, can ever achieve a good society. Instead, such practice reintroduces the very dilemma critical praxis seeks to overcome.

Critical Theory's Response: The Teleological Moment

Now, both Honneth and the later Habermas recognize this problem, and both work out conditions under which the dilemma may be resolved. The problem is that in order to bridge the gap between the individualist conception of emancipation and the social practice necessary to both developing agency and moving toward a good society, both critics introduce certain "teleological moments" into their notions of critical theory.[59] The effect is to render their versions of critical theory a bit less thoroughly critical. Habermas, for example, posits certain shared interests among "free," unencumbered agents that impel them first toward "communicative action" and then toward an ethics of discourse that is an "unavoidable presupposition of . . . argumentative practice . . . forcing itself intuitively upon anyone who is at all open to this reflective form of communicative action."[60] In this way, Habermas sneaks teleology in through the back door. Positing shared interests at the outset is arguably a form of ideological domination and arguably arrests critique of the shared interests and resultant ethics. In any event, his posited "shared interests" and "ethics of discourse" seem to be in the nature of abstract, universalizing assumptions that ignore the concrete practices of people as situated in real-life contexts. Consequently, for Habermas, a good society does not really seem to involve emancipated individuals working out shared interests and an ethics for themselves in an ongoing fashion. Rather, he seems primarily concerned with articulating the intrinsic desirability of some proffered ideal, the very thing that critical theory originally counted as the root of the dilemma. More recently, Habermas has recognized these problems, holding that his "universalistic principles" of discourse have "merely presumptive generality."[61] Consequently, they do not necessarily resolve the dilemma.

Honneth and Farrell suggest that the resolution resides in "those circumstances experienced as unjust . . . on the basis of the criteria that affected subjects themselves use to distinguish between a moral misdeed and mere ill luck."[62] This individual experience of injustice is linked to institutions both within and without the "lifeworld" (e.g., the family and the law), and granting each person recognition and respect in the institutions of both the "lifeworld" and the society is linked to emancipation and the realization of a good society.[63] As in Habermas, however, teleology is introduced. People should behave in certain ways in order to effect the development of emancipated individuals. But what if acting that way does not accord with the chosen values and lifestyles of the emancipated individuals so produced? Again, individuals are not left to work things out for themselves, but are urged to follow an intrinsically desirable proffered ideal.

Moreover, in Honneth's version, the development of a good society seems more dependent upon institutional than individual choice. Why bother with emancipating individuals if the realization of their identity interests and emotional security may be effected from the top down through institutional manipulation (i.e., recognition) rather than individual choice? Arguably, at least, an ideal authoritarian government could act as a corrective for the problem of those unemancipated by either the communicative power of ideal speech situations or because of the denigration experienced within familial, civil, or legal institutions. Similarly, coercive power employed to meet and legitimate both shared and unshared interests and practices would seem as effective at emancipation as communicative power or respect. Under certain circumstances (e.g., broadly shared extreme racial bias), it may even seem necessary (e.g., forced busing to achieve school integration). Arguably, such problems are remediable through courts, continued sincere discourse, or respect in a range of social interactions, as Habermas and Honneth suggest (though it seems that courts might need to resort to coercion as a final remedy), but this seems an act of faith engendered by an ultimate dedication not to emancipation but to a particular ideal societal and governmental form.

The problem is that both Habermas and Honneth seek to ground their notions of what constitutes emancipation and what a good society does somewhere beyond the mundane, and this maintains rather than dissolves the dilemma. By including universal presuppositions of communication and normative presuppositions of social interaction in their versions of critical theory, Habermas and Honneth seek to retain an individualist notion of emancipation while moving toward a nonindividualist (shared) notion of a good society. However, both sets of presuppositions are "disconnects" or leaps from things as they are to things as the critics think they should be. This tends to obscure the reality of our living together in concrete situations and sorting out how we can "go on" just at the moment, and continues the theory-practice gap that is at the dilemma's core. Perhaps our concern should be not with a priori communicative principles or a priori norms of social interaction but with how we routinely react and respond empirically to the tendencies in each other's conduct as revealed in contextual practices mediated by institutions of all sorts. This "shifting substratum" of interpersonal behavior within the background context of institutions is arguably the actual ground of identity and emancipation as well as the point at which we begin constructing alternative, mutually acceptable ways of going on toward a good society. What is needed, in other words, may be a thoroughly nonteleological critical theory that understands emancipation in a different way.

Evolutionary Critical Theory and a Good Society

It is our contention that both the return of the dilemma and the failure to resolve it with "teleological moments" arise from critical theory's individualist conceptualization of emancipation. Furthermore, a possible solution to the dilemma may derive from scholarship suggesting a more thoroughly nonteleological version of critical theory (Evolutionary Critical Theory) that understands emancipation in a different way. Evolutionary Critical Theory acts as a conceptual bridge, allowing us to move beyond many of the constructed limitations that bind traditional critical theory and hinder its application. A fundamental aspect of Evolutionary Critical Theory is the perception that we are never disconnected from power relationships and their mediating institutions. In addition, this connection is not unreservedly adverse to our interests or our emancipation.[64] This concept is grounded in Veblen's understanding of "usufruct" and how it articulates the power relationships between the public and private spheres.[65] He further articulates an understanding of society that "is a joint possession of the community" where control "has come to be effectually vested in a relatively small number of persons."[66] Therefore, we can become more thoroughly involved (through both positive and negative critique) in the power relationships and institutions about us rather than rejecting the dominant ideology, its discourses, practices, and institutions by stepping outside of society to seek the good society. This, in turn, allows us to retrieve critical theory from Horkheimer's pessimism, Habermas's reliance on "ideal" situations, and Honneth's "teleological moments" by fusing theory and practice (a consistent goal among critical theorists) at the concrete, moment-to-moment, indeterminate, intersubjective interaction of individuals in institutionally mediated contexts.

These intersubjective relationships, therefore allow us to craft a means to recognize the moments, the adaptive practices within them, and the institutions thrown up out of them, not as the echoes of an overarching total domination, but as the very stuff of emancipation and the pursuit of a good society. It thus avoids theory-practice gaps at the source of the dilemma by eschewing all teleology.

This does not by any means imply that there is a total dissolution of the principles of liberty or individuality. All of the groups, individuals, and actors are, in fact, vested with the inalienable right of self-determination. Veblen, unlike Horkheimer, Habermas, Honneth, and others, eschews the teleological elements of "high-minded" motives, stating that such self-determination would often be understood as the means of "self aggrandizement of each and several at the cost of the rest by a reasonable use of force and fraud as the traffic will bear, in this national enterprise of self aggrandizement."[67] Though it would appear that Veblen has completely eschewed the emancipation project at this point, in reality he lays the groundwork for how emancipation, cooperation, and other societal goals are achieved. He states in regard to the right of self-aggrandizement that "the nations, each and several, can continue to exercise these rights only on the basis of a mutual agreement to give up so much of their national pretensions as are patently incompatible with the common good."[68]

In this process of negotiating a good society, these rigid "institutions of domination" are in reality quite fluid, since the individuality, the sovereign "rights" of these institutions, can be maintained only in a fragmented state.[69] In essence, the lack of "better" motives creates a system that fosters intersubjective relationships, thereby moving actors collectively toward the sort of "microemancipation" proffered by Alvesson and Wilmott, which operates within organizational systems rather than against them.[70]

As a basis for such an argument, Veblen begins with a critique of enlightenment thought as embodied in classical and neoclassical economic thinking. In both *The Theory of the Business Enterprise* and "Why Is Economics Not an Evolutionary Science?" Veblen objects to the classical notion that human behavior (economic or otherwise) is subject to "laws" or "final causes."[71] Pointing out that in reality we must simply contend with a confusion of "brute facts,"[72] he argues that the difference between theories about what these facts mean "is a difference of spiritual attitude or point of view . . . or in the interest from which the facts are appreciated."[73] More succinctly, "it is mainly a difference in using different frames and preconceptions to organize facts in a meaningful structure."[74] There is, in other words, no epistemological, communicative, or normative advantage to choosing one theory over another. It is simply a matter of choosing which habits of thought best serve certain interests and goals or which helpfully describe what is going on as interpersonal drift is built up into fluid patterns through emulation, habit, and institutionalization.

Thus, at the outset, Veblen eschews both teleological moments and all concern with articulating the intrinsic desirability of some proffered ideal. In contrast, he offers simply a pragmatic, prolific, undirected trial and error. So long as this trial and error is unconstrained artificially (e.g., by human institutions), everything that can happen will happen; and all that can find purchase will remain (at least for now). Thus, it is a historical and "evolutionary" theory without trend or consummation. Behavior is explicated in terms of the day-by-day adaptation of people and institutions to actual events of immediate importance to extant individuals, and change is elucidated as a process of choosing whatever forms of immediate behavior within the background context of social institutions (whether thought "progressive" or "regressive" in theoretical terms) that "most fit" the current historical happenstance. Institutions are accounted as nothing more than habits and tacit agreements about how to proceed in that current environment (social, political, and natural), and these habits and agreements can not only be changed consciously but are in fact modified moment-to-moment as we work out how to go on just now. Additionally, this evolution does not proceed through a selection of traits by the external environment. Rather, it emerges from internal variation through the sort of endogenous relational practices just mentioned and resulting in "drift," cumulative emulation, habituation, and institutionalization. External influences certainly impact the system, but they must be included in the set of relational practices and discourses operating and drifting in the system in order to matter to people. Moreover, conscious change occurs only when an understanding of those external influences is sufficiently worked into the dominant ideologies, practices, and discourses of the system,[75] and only when those within the system realize that their shared social realities are dynamic.[76]

Veblen's *Theory of the Leisure Class* develops and deepens this argument.[77] Therein he contends that certain unintended ills arise out of the habits and tacit agreements. First, the agreements usually privilege some people. Second, some (e.g., the "leisure class"), in pursuit of these privileges, work to inhibit the "natural selection" of institutional forms better suited to the historically emerging social and economic environment. This inhibition is reinforced by both the propensities for emulation by others and the habitual patterns of thought ingrained by current institutional arrangements. Thus, throughout Veblen's thought is a deep appreciation for the synergy among forms or modes of living, individual patterns of thought, institutional structure, and community organization.

Particularly, such individual schemes of thought are reverberations or versions of socially developed life schemes. These schemes, in turn, could arguably lead to fragmentation. However, there is a synergy among the movements that ultimately brings a community into some "passable consistency" through legal and moral rights and obligations.[78] Concerning this synergy, Veblen finds instances of maladaptation particularly interesting. When habits of thought lack congruity with the historical economic setting (with, e.g., the occupational structure or available technologies), distortions occur in the evolutionary process. Some distortion is "natural" because institutions are adapted to prior stages in the evolutionary process and thus never fully accord with the immediate situation. Some distortion, however, is "sociological." It is occasioned by vested interests fending off change, the persistence of inappropriate patterns of thought, and the attractiveness of occluding metaphors ultimately leading to evolutionary "dead ends."

Most important for our purposes, to Veblen all our experiences with the synergy among institutions, individual patterns of thought, institutions, community organizations, and the maladaptations wrought by natural and sociological distortion, along with their multiple effects through life, shape and affirm our identity and capability as agents. On the one hand, we react and adjust to the effects and the "feedback" of reactions they engender; but on the other hand, we affirm our "selves." We define our boundaries synergistically within these structures and relationships, but we also put others on notice and engender an identity in the lives of others (individuals, groups and institutions). Throughout his "Introductory" to *The Theory of the Leisure Class,* for example, Veblen describes the many forms of unfolding taken by the individual agent in the contexts of interpersonal, discursive, and institutional relationships.[79]

What Veblen would have us include in our critical analyses, then, are these nondeliberative "background activities," the things we just do and often just do a little differently in order to be doing "the same thing" in different situations and institutional contexts, the things we do not so much in order to pursue our interests (real or socially constructed), but to engage the social and institutional circumstances of our context at the moment. For Veblen, the Frankfurt school's domination and emancipation, Habermas's "communicative power," and Honneth's norms of social interaction are thrown up from and make sense only within the ongoing stream of social life from within which they arise and within which they have their application. Hence, there is nothing unavoidable or morally necessary about them.

Moreover, if we appreciate the synergy among individuals, agency, institutions, and the social system at large, we then avoid the category mistake of "deducing the larger social forms from the smaller" and building up "notions and models of the collective from primary accounts of the individual actions and immediate face-to-face encounters."[80] Veblen recognizes that the whole is greater than the sum of the parts and that it sometimes turns around on those parts, oppressing them by distorting or precluding experimentation. He points out only that the institutions that are part of that greater whole may also be employed to emancipate. Whether they are emancipating or oppressing, he suggests, may be revealed by including immediate, face-to-face encounters and individual action in our critical analysis not only as the result of domination mediated by institutions but as the stuff that throws up the greater whole as well.

Implications for Critical Theory

For our purposes, the most important thing that Evolutionary Critical Theory means for critical theory generally is that because our capacity for agency (emancipation) is born of the relationships and practices we encounter moment-to-moment, they precede individual existence, desire, choice, and free action, not vice versa. Thus, change in the meaning of a good society and what we do as agents to pursue it are the results of endogenous forces operating through the relational practices, discourses, and institutions of the society. Perhaps they are the reactions to the truths revealed to us by the "multiple constraints" of our norms and institutions.[81] Perhaps they follow from internalized exogenous forces. Perhaps they result from the constant drift in the substratum of forces exerted everywhere within the social structure.[82] In any event, the contexts of our lives have a form molded by the dominant discourses, relational practices, and institutions of the society; and getting the good society right or successfully emancipating cannot be accomplished outside of that form.[83] There is simply no "place" to go save another "form."

For example, if I say that I'm oppressed and desire emancipation, I am giving my subjective psychological experience an objective reality by categorizing it under certain already established verbal expressions. Other people may now understand the sort of experience I'm having because I have brought it within a certain "language game" that has certain shared forms molding the right and wrong uses of terms. These common forms and language games were relationally constructed before my experience and are available for my use in perceiving the character of the reality around me. Moreover, it is difficult to imagine how similar experiences (giving rise to similar interests) enable communication if there is no extant language (construction) enabling us to categorize our experiences under the same form. Of course, available expressions and language games may not fully capture my subjective experience and may even denigrate and work against my understanding; but to move toward a fuller capture I must begin with the closest thing to hand in order to meaningfully engage others and even myself in the critique of what is available and oppositional so as to begin the movement to something else.

As I try to determine whether a particular practice is conducive to fuller capture or to emancipation, or even in accord with what would be to me a good society, the determination of what counts as "in accord" and what counts as "conducive" is necessarily made ("negotiated") at the moment, within a particular concrete circumstance with attendant, ongoing social practices. Community norms at that moment may, in practice, inhibit change and stabilize patterns of "disadvantage" and "oppression" in the name of preserving or advancing a good society. However, they also reveal those patterns as "truths" about the community and its norms; and once these truths are experienced in practice, they may lead us to reconstruct our norms and our notion of a good society. As Foucault argues, "Truth is not the reward of free spirits . . . [it is] a thing of this world . . . produced only by virtue of multiple constraints."[84] For example, we neither know how imprisonment works nor understand its effects until we imprison. Thus, we do not know how we will evaluate it before our experience with it. Similarly, we do not know the nature or extent of the emancipation or oppression a set of community norms may provide until we establish and live with them. Though we may analogize and employ metaphor prior to experience in order to imagine what our evaluations might be, nothing substitutes for experience. Hence, the truths we know depend upon the patterns of practices constituting our society's "regime of truth," and what we mean by "emancipation" and a good society is similarly dependent yet flexible and open to "drift" and change.[85]

Put another way, we must think or believe that something is true before we can choose (freely) to act accordingly. Beliefs arise from many sources but, most broadly, always from either some form of authority or some form of direct experience. Most of critical theory seeks to reject all authority and experience that is mediated by authority and to test choices by standards developed by an unencumbered, uncontradicted, self-defined individual.

However, this is impossible. No such creature can exist given the intersubjectivity of self-identity and the fact that all that we know "is produced by virtue of multiple constraints" and the experiences they engender.[86] It is only through active participation with and within the practices and habits of thought of our society and its institutions that we gain the experiences necessary to forming beliefs contrary to those authoritatively impressed. Effective individuality (emancipation) is thus not individualist but reliant. It relies upon our synergy with everything about us; and the more we engage (negatively and positively) with all about us (including our institutions), the more opportunity we have for individuality. It is through just such engagement that we individuate.

There are other important implications for critical theory as well. First, Evolutionary Critical Theory suggests that individual disengagement and negative critique alone preclude any appeal to some intersubjective notion or coming together on what counts as emancipation. Moreover, because the very sense of being an (emancipated) individual agent living in a good society derives from the total daily experience of the interchanges within such relationships, the more disconnected (emancipated) we become, the more indeterminate becomes our emancipation or sense of direction toward a good society. In fact, any notion of a good society is impossible, as disconnected, negatively critiquing individuals cannot put to issue or address any standard of community interest in our individualist discourse. In brief, speaking of a society of emancipated individuals as opposed to one based on "force" or "power" involves claiming that society should be structured and should function in a certain way. That, in turn, involves appealing to something other than personal or group interest, and that requires positive critique, social engagement, and institutional membership. This means that emancipation and a good society cannot involve moving from the pressures of engagement in power relationships and their mediating institutions to an unencumbered freedom, but instead must involve moving from one set of power relationships and institutions to others that we perceive as enhancing our ability to satisfy needs and interests as experienced within the current context. Political, social, and economic institutions, within the Evolutionary Critical Theory framework, can provide opportunities rather than only hindrance and oppression for those actively engaged in constituting themselves as agents and acting in pursuit of a good society.

Second, the popular idea among some critical theorists that there are private, individual notions of a good society that need to be included in the discourse on social practices and institutional forms (that "your good society is not my good society") is epistemologically problematic. This individualist approach is founded upon the idea that because I cannot know whether your experiences are like my own in similar circumstances, I cannot know what you mean by "emancipation" or a "good society" when you employ the word. So, for all we know, each of us may mean a completely different sort of thing. By implication, of course, if we are to accomplish a good society or emancipation, we must address each person individually and meet his or her definition. In a sense, this idea obliterates the concepts of emancipation and a good society by reducing them to impossibility. If my understanding of your good society depends upon my having precisely your experiences, your notion of the term is forever lost to me. Evolutionary Critical Theory avoids this implication by stressing that meaning is socially negotiated, largely within the contexts of positive and negative institutional critique.

The final implication that we would like to mention here is that although Evolutionary Critical Theory recognizes our ability to invent all kinds of theories as to what emancipation or a good society is (e.g., egalitarian, utilitarian, libertarian, hierarchical, highly networked anarchism), no theory can ever be a final account. Many critical theorists are convinced otherwise. They are convinced that the tradition of Hegel, Marx, the Frankfurt school, Habermas, and Honneth discovered and continues to reveal the essential meaning of "emancipation" and therefore of a "good society" in a dialectical tradition currently expressed in terms of an ongoing individualist discourse or social practice conducted according to certain ethical norms. Evolutionary Critical Theory, however, maintains that all such theories and traditions are derived by us and "live" only within the dynamic context of social life from within which they arise and within which they have their application. They cannot be turned around to depict or portray the dynamics themselves. Consequently, the meaning of a good society depends upon our form of life, and as that changes so does the nature (the meaning, the essence) of a good society and what we must do as agents to secure it. The meaning or nature of a good society is in this sense indeterminate; and while its meaning is neither a matter of strict logic nor practical necessity, it is not arbitrary either. It is socially and humanly coherent. It is not essentially any particular thing, but it cannot be just anything at all. The meaning of a good society is coherent exactly to the extent that our form of life and its transformations are coherent. In brief, it is evolutionary. In this sense, Evolutionary Critical Theory avoids the nihilism and rampant normative relativity that to a certain extent motivates Habermas's and Honneth's search for necessary intersubjective communicative and normative presuppositions.

Conclusion

If our understanding of Evolutionary Critical Theory is correct, a good society can have meaning and be pursued only so long as there are right and wrong ways of using the term, and the right and wrong ways cannot be privately or individualistically determined. Rather, they must be collectively developed and continually negotiated within the context of power relationships and their mediating institutions as we go about doing things with each other on a moment-to-moment basis. To know the meaning of the good society, to critique it, and to suggest alternatives and directions to take toward it requires that we know how and in what kinds of situations or states of affairs it is used effectively for shared purposes in our community. Thus, the meaning of a good society must be socially coherent; and within much of critical theory's individualist discourse, as it posits no community of interest other than aggregated individual interest, a good society is meaningless.

What is most interesting for our purposes is that this reconstruction of critical theory both solves the dilemma engendered by the fact that respecting individual value and lifestyle choices threatens to frustrate the realization of a good society and suggests a legitimate role for institutions in resolving this dilemma. Evolutionary Critical Theory resolves the dilemma by both agreeing with most critical theorists that it is more theoretical than real (being engendered primarily by the idea that some proffered ideal society is intrinsically desirable) and by insisting that the dilemma can be resolved only in praxis and not through universal presuppositions of communication or normative presuppositions of social interaction. Rather, people must work the apparent contradiction out for themselves through socially coherent praxis, thereby closing all theory-practice gaps. Institutions, in turn, are legitimate to the extent that they empower individuals (not individualistically but "reliantly" and synergistically) to act as agents while creating a "space" for the synergy of intersubjective experiences that move these societal processes. Toward this end, legitimate institutions aid the trial-and-error process by responsively putting ideas into practice, thereby revealing the positive and negative implications for society, groups, and individuals à la Foucault. By operating in the context of mediating institutions, Evolutionary Critical Theory promotes first the process of microemancipation and then the process of evolutionary change.[87] By increasing involvement in praxis rather than increasing detachment and isolation, individuals become increasingly aware ("reliantly" and synergistically) of their true individual interests as they come to understand them in practice. This, in turn, places these individuals at the heart of the change process, effectively empowering all actors within the societal system.

As a result, these mediating institutions must have some logical framework drawn from both theory and praxis to govern the process. Societal existence thus requires an entity such as public administration, with its specific ontology, intersubjectivity, and inherently critical nature, to manage and cull power flows and relationships, negotiate the vested interests of the jointly held body of a society,[88] distribute societal emoluments, and foster our movement toward a good society. Public administration, therefore, by virtue of its theory and praxis, is both *evolutionary* and *critical,* providing the conceptual and public space that allows for the sort of "endogenous evolution" we have described throughout this piece, while facilitating movement toward more "fit" institutions. As a final consequence, public administration becomes responsible to and representative of *every* facet of society because it embodies the space for a synergy among institutionalism, hermeneutics, and traditional social science while fostering the intersubjective experience of good governance in theory and praxis. This allows us to reflect on a few central issues in public administration, including identity, legitimacy, and epistemology, using Evolutionary Critical Theory.

8

Evolutionary Critical Theory and Public Administration

The idea that public administration might exist outside of a mother discipline, with an ontologically sound disciplinary matrix, tends to be a matter of continued concern for public administration. How then could public administration function legitimately if it is not linked to some aspect of the public sphere, within a democratic society? Such concerns have been raised by Rohr, Terry, and others,[1] typically with scholars attempting to shoehorn public administration into a series of roles often while maintaining its subservience to some elected body. However, if we return to our thesis that public administration *synergistically* employs the methodologies of traditional social science, institutionalism, and hermeneutics by using a reconstructed critical theory as synergistic agent, we might offer a more provocative idea to consider, that public administration functions in theory and praxis as the center of synergy in a society. This function effectively crafts a unique identity, a unique source of legitimacy, while reconciling what many would consider a fragmented epistemology using Evolutionary Critical Theory.

Throughout this book, we have emphasized that public administration is a unique discipline. It necessarily requires us to understand the complexity that surrounds this intersubjective experience of good governance at its core, creating a provocative foil to the notion that theory is dead and that the cultural relativism of situationally[2] derived "islands of theory"[3] is not absolute, but instead coheres around the very real unit of intersubjective experience. This, in turn, allows us to understand that public administration is a discipline with ontological status while realizing that continued debate around arguably central questions such as legitimacy, identity, and epistemology reflect the processes of endogenous evolution taking place as the disciplinary matrix negotiates them in or out of its customs and practices. Such a process addresses how public administration copes with "day-to-day routines . . . and the all-inclusive speculations comprising a master conceptual scheme."[4] Consequently, we were then able to demonstrate how evolutionary conceptions of power, emancipation, and the good society both provide and require the sort of public space that public administration provides, leading us to the concluding view that many of the central debates that have rent the discipline over the years can be understood and possibly reconciled using Evolutionary Critical Theory.

It is important to recall that in Evolutionary Critical Theory, goals such as emancipation, integrity, and equality lose their ideological status, and instead gain their meaning from the intersubjective experiences that the context of the moment provides. Achieving such ends then becomes hinged to the urgency one places on them and the resources that are available and used for these ends. As a result, the theory-praxis gap closes, since societies either adapt and change (adopting fitter institutions and processes), or stagnate and die. These adaptive processes *must* be guided by immanent critique both of discourse and ideology, focused on the language games, extant life, and intersubjective processes of a society, leading us toward understanding seemingly irreconcilable issues, such as power, emancipation, and the good society, that have consistently vexed other scholars of critical theory.

Traditional critical theory has consistently been unable to cope with issues of agency from within an administrative state and has been particularly unsuccessful when attempting to negotiate "positive" outcomes like unencumbered agency within a "less than minimalist" administrative state. In this sense, the most important thing that Evolutionary Critical Theory does for critical theory is that it focuses our capacity for agency (emancipation) on the relationships and practices we encounter moment-to-moment, which precede individual existence, desire, choice, and free action, not vice versa. Therefore, a good society and what we do as agents to pursue it grows from endogenous forces operating through societal practices, discourses, and institutions. They can be reactions to insights about our norms and institutions, including their constraints.[5] They might grow from "internalized" exogenous forces, or possibly from the constant "drift" of forces exerted everywhere within the social structure.[6] In any event, our lives are molded by these dominant discourses, relational practices, and institutions of the society. Creating a good society or emancipating it cannot happen outside such a form.[7]

Evolutionary critical theory both solves the dilemma of individuality and suggests a legitimate role for institutions in resolution process. Such a resolution occurs when we suggest that an ideal society is theoretically desirable and that its achievement can happen only when people work out the contradictions in their current scheme in a socially coherent praxis, instead of through universal norms for communication. This creates a space for intersubjective experiences that in turn drive our social processes from within mediating institutions, promoting both microemancipation and evolutionary change.[8]

When we increase involvement in praxis rather than detaching or isolating individuals, they grow to understand their true interests in practice. This in turn makes individuals central to any change, ultimately empowering everyone to move through these logical frameworks of institutions, creating a joint conception of a good society.[9]

Now, we can begin to address the three contested issues central to debates concerning the nature, roles, and responsibilities of public administration. Within Evolutionary Critical Theory, we can reflect on debates about identity, legitimacy, and epistemology, allowing us to contextualize them, sympathetic, erroneous, or incomplete, based on what we now know about the discipline of public administration. Therefore, understanding Evolutionary Critical Theory allows us to fully apprehend the complexity and richness of public administrations theory and praxis. Specifically, in the following sections, we can argue that public administration is a unique, ontologically grounded discipline that is necessarily a creature of society and not one of a single state or organizational entity.

Identity

We have established public administration as a discipline with ontological status. This, in turn, has several implications for the understanding of its identity. Identity, in this sense, involves our understanding of what public administration is and where it fits among other disciplines. Although research on the professional identity of public administration has been pervasive, it consistently falls short when trying to encompass its totality. Some of the earliest modern scholars justified the practice under the rubric of science and, later, scientific management.[10] Over time, perceptions changed and scholars began to reexamine the roles political theory might play in the theory and practice of public administration.[11] Similarly, others have examined how public administration relates to other disciplines, including business administration[12] and political science, more broadly.[13] Such a disparate body of research alludes to the necessity of understanding how public administration relates to management, political science, and business administration, both in theory and in practice.

We demonstrated in earlier chapters that public administration is a discipline unto itself with ontological status. This challenges assertions that it is not simply a subfield of politics or political science,[14] nor is it a subfield of business,[15] warranting some type of generic administration. However, since public administration has a disciplinary matrix with a clear ontology, it can, as a result, negotiate roles for the variety of disparate fields, allowing them to meaningfully contribute elements of theory, methodology, tools, and techniques. So even though public administration is not management, nor is it political science, or even a field made up of a complex of disciplines, the disciplinary matrix of public administration includes spaces for public administration as "public management,"[16] as "administration,"[17] political theory,[18] and even its "interdisciplinary" nature.[19] Such a situation makes it useful to briefly elucidate how each issue fits within our argument.

Broadly, many of the arguments that attempt to ground public administration's theory and practice in one discipline or another come from similar sets of circumstances. Typically, change and development in public administration have tended to cohere around broad societal issues, such as the Progressive Era, the rise of scientific management, and the behavioral science movement. Later, some of the foci changed to include developments from the human relations school, economics (privatization and policy analysis), and others. Two disciplines in particular, political science and business administration, have consistently informed the study of public administration, while consistently considering it to be a subfield within their broader scope of influence.

One might argue that a catalyst for public administration as a field of business came from the Progressive Era's reaction to corrupt business and public practices in the late 1800s. The consequence of this corruption in both the public and private sectors was the belief that the practice of politics was corrupt and could be corrected only through the judicious application of the scientific method to the process, giving rise to "scientific management" in both sectors. This belief evolved over the following decades, adopting many of the tools and methods of the behavioral sciences, the movement toward privatization, and its most recent incarnation, the "new public management," which has sought to reincorporate some of the more positive aspects of the public element of public administration, while retaining its core teleological beliefs.

As a consequence, public administration bears some family resemblances to business administration, including many of the tools and methods of management and leadership. However, the teleological assumptions about goals, processes, and responsibility central to the study and practice of business administration remain functionally incompatible with the study and practice of public administration, making the two quite incompatible. Despite such incompatibility, if we choose to emphasize the administrative aspect of public administration, scholars find a relatively powerful, yet incomplete justification for its existence and a number of resources for the tactical or street-level situations administrators face, although the incompatibility of goals alludes to the reasons why business administration cannot encompass the disciplinary matrix of public administration either in theory or in praxis.

Having a historical tie to both the public sphere and political theory makes the argument for public administration as a subfield of political science rather compelling, despite a set of incompatible goals, processes, and responsibilities. Such incompatibilities reflect issues we highlighted in our earlier discussion of business administration. Intuitively, political science links rather closely to the public aspect of public administration.[20] At the core of such arguments, we find some historic context about the Constitution and its framers,[21] arguments for public administration as a creature of the state,[22] and applications of political theory.[23] Although these arguments have a great deal of merit, in many cases becoming a part of the negotiated core of the discipline, they do not represent its totality, leading to more critique and debate rather than illuminating public administration's public role.

Most recently, however, some scholars of political science have moved away from the well-grounded arguments expressed above toward a position, arguably far weaker, grounded in instrumental rationality, where political science acts as a "conversion variable."[24] These scholars have been criticized for seeking a "simplistic political system shorn of troubling normative attributes not amenable to their brand of scientific analysis. Needless to say, such a view hardly comports with the governance of a complex society in a global economy."[25] However, Whicker, Strickland, and Olshfski must implicitly realize the limits of tying public administration too closely to political science, since they call for an "interface between the two disciplines."[26] This points toward the understanding that the best they can hope for is that political science might inform the study of public administration in some way, but cannot claim or reclaim it as a subfield. The debate about public administration and political science continues, though other approaches tend to include statements that, for example, such as the "intellectual space of public administration has been highly defined" by the "scientific politics" project with ontology and theories constituted by politics as a "trusted science."[27] Even in this argument, Lee acknowledges that this might not be a one-way "dominating" relationship, but instead might involve the two engaging as equals.[28] In these cases, such a realization represents a movement toward a shared dialogue. Such a dialogue then enables political science to inform *some* of the strategic issues faced by the discipline when we emphasize the public aspect of public administration, painting a useful, albeit incomplete picture. Over time, as the discipline of public administration continues to mature and change, the argument that it fits as a subfield of political science becomes far less provocative. This argument becomes diminished further when its proponents continue to offer one of the weakest justifications, that of a single approach to inquiry,[29] which remains wholly inadequate for understanding the synergy among the intersubjective experiences of good governance.

Legitimacy

Public administration, particularly in the United States, has continually attempted to justify its existence within any number of the constitutional frameworks, including the executive, legislative, and judicial branches; the government as a whole; and even from a technical approach. Given that we understand public administration is a discipline of intersubjective experiences, operating in the context of Evolutionary Critical Theory, we can now understand how teleological beliefs often make notable authors unable or unwilling to pursue a notion of legitimacy societally rather than organizationally. Public administration's orthodoxy approaches legitimacy from three basic areas: those who believe it is legitimate based on its ties to the presidency, to Congress, or to the judiciary. In addition, other scholars have wrestled with this legitimacy issue from more complex positions. Though a number of scholars have built their careers supporting or railing against such conceptions, we intend to present the basics of each argument and systematically question our understanding of them using Evolutionary Critical Theory.

The Judiciary

One of the most vocal, if not persuasive scholars who advocate public administration's place under judicial control is Theodore Lowi.[30] He argues that administration is necessary and is a part of our capitalist society even though it is not a part of capitalist ideology. However, like many other scholars, he is convinced that public administration, unfettered, can easily degenerate into a cadre of unresponsive, overly powerful administrators[31] who would tend toward Fascism.[32] Lowi uses this context to frame his basic argument that public administration must be tied explicitly to the judiciary in order to prevent the abuse of power and act as a mechanism to provide controls over capitalism in America. This framework then will foster the "liberalism" among individuals necessary to prevent the unraveling of our republican form of government. According to Lowi, such an end can be achieved only through a judicial solution because "interest group liberalism demoralizes government, because liberal governments cannot achieve justice."[33] He further argues that interest group liberalism "corrupts democratic government" by weakening its capacity "to live by democratic formalisms."[34] This, in turn, incapacitates society's ability to allocate responsibility.[35] These effects, according to Lowi, unhinge the structure of our government,[36] potentially leading a substantial minority of people to reject democracy in favor of some "moral republic in which basic values are so homogenized that democracy can take place without risk of morally bad outcomes."[37]

Lowi, much like traditional critical theorists, falls prey to the limitations of teleology. This "republican era" is a constructed political concept, an artifact of our current society that may or may not lead to the doomed end state envisioned by Lowi. However, it is intriguing to note that Lowi and many other critical theorists become "melancholy,"[38] turning toward pessimism as a consequence of their teleology.[39] Goodsell offers a counterargument: "if the conservative coalition does not fly apart as Lowi predicts and instead moves ahead to rule over the new century, then public administration can provide many counterweights to safeguard the republic."[40] It appears that Goodsell, unlike Lowi, has a tactile understanding of public administration's potential to foster the intersubjective experiences of good government.

Such an understanding suggests that, from an Evolutionary Critical Theory perspective, we might move toward a "fitter" institution driven by our intersubjective experiences, or the "framework of operative norms" essential to protecting responsible governance.[41] If this does not happen, our society might stagnate or possibly crumble. Lowi also falls prey to what can be understood as a "paradox of responsibility," which occurs when someone tries to use a small part of the whole instead of the negotiated whole. This, in turn, can lead to the ills of interest group politics that Lowi fears most. If we understand that public administration is responsible to the society writ large, and not necessarily to any single local interest, then the society, when faced with Lowi's fears, will respond with the endogenous evolution that is central to adaptation in Evolutionary Critical Theory. Over time, these concerns become possible mechanisms for drift as we move toward "fitter" constructs. These ideas of "liberalism" and the "republican era" as socially negotiated concepts might change as society changes, resulting from what many scholars might construe as the current and changing "will of the people."

The Legislature

Lowi is by no means alone in his quest to "capture" public administration within a single branch of government. An even more vocal Schoenbrod, for example, has made a career of challenging the delegation of power.[42] He makes a number of arguments to demonstrate that agencies reflect "concentrated interests" instead of the will of the people. He points to the process of lawmaking and how it requires expensive representation and strong ties to members of Congress, which pressure agencies into the service of these "concentrated interests."[43] Schoenbrod would have Congress cease passing the vague rules that are the ultimate result of large-group decision making[44] and would force it to relinquish the power and image enhancements that delegation provides[45] despite the historical problems associated with implementing it, including its failure in the judiciary.[46]

These power and image enhancements are central to congressional politics. They allow politicians to support popular legislation such as clean air, while shifting the blame for costs, implementation, and monitoring to bureaus and agencies.[47] Schoenbrod's solution to these problems, of course, is to remove the ability to delegate from Congress and force it to make "clear" laws that would not require interpretation, delegation, or even political negotiation. This is of particular concern to him since he believes that "delegation makes it easier for government to infringe upon liberty" while it is "unlikely to result in more effective protection of the public."[48] He states further that democracy is "meaningless because it is impossible to be certain that [decisions] are not simply an artifact of the decision process that has been used."[49]

In this sense, Schoenbrod echoes the pessimism of both Lowi and traditional critical theorists by falling into the traps of teleology and the paradox of responsibility. In Schoenbrod's case, public administration is not truly to blame for the ills of society brought forth in his arguments. The blame instead falls on the flawed but very real methods Congress uses to conduct politics. This flawed process, however, is a part of the fittest institution *at that moment*. If a group such as Congress systematically demonstrates an inability to provide space for good governance, then arguably it will adapt, fail, or die. If anything, such a situation allows for the type of broader but controlled delegation proposed by Spicer and others.[50] Once people in their negotiated relationships question who should be entrusted with control over a discipline and practice that holds the intersubjective experience of good governance, then the process of evolution and change could begin reshaping praxis. It could, indeed, turn our focus toward a societal responsibility (i.e., to a good society) where administrators in practice might be, over time, more inclined to support its position and its role, rather than bowing to the whims of political fancy. Thus, it could lead to a fitter conception of governance.

The Executive

One final group of scholars attempt to place public administration exclusively within the executive branch. Jones argues that public administration fits into his understanding of a "diffused responsibility" perspective.[51] From such a perspective, the adaptation and drift toward fitter institutions that we discuss throughout this book would be difficult, since governmental structure is "antithetical to efficient goal achievement."[52] In addition, such a role in practice plays into common presidential behaviors, including the process by which a "president and his associates is to plug into that permanent government as a prerequisite to establishing a degree of influence or control."[53] Consistent with other literature on the presidency,[54] Jones argues that administrators understand these power and influence games quite well, and often comply. Jones, unlike Lowi, Schoenbrod, and others, realizes that this process is "temporary" and a part of an "ever-changing organizational design."[55] In addition, presidents often lack sophistication in the process of establishing control, often relying on trial and error. This, in turn, can lead to the appointment of a vast body of administrators with "little or no background or training in the organization and management of the federal government."[56]

This process of diffused responsibility illustrates how functions and organization can change to accommodate goals within a political system. "A White House staff of close advisers and a presidential branch have emerged over the postwar period to enhance the president's control, but these developments have created as many problems as they have solved."[57] Since this lack of sophistication leads to the choice of greater control at the expense of skill, presidents often find themselves at the mercy of professional rather than the political challenges wrought from this process. According to Jones, the process remains authentic, be it "split-party" or "single party"; if analysts and the public believe they are equally legitimate, then in practice both forms could "work well," effectively demonstrating how some intersubjective experiences are negotiated in practice.[58] Though his conception of public administration is incomplete, Jones does not appear to share the same level of pessimism as other critics. Instead, he, much like Goodsell, appears to have some tactile understanding of the endogenous evolutionary nature of governance.

Other Approaches

A number of scholars, including Rohr, Terry, and Spicer, have taken a different approach to this dilemma of legitimacy. They either explicitly or implicitly attempt to legitimate public administration as a part of the whole. Each has found it problematic or at least difficult to capture the intersubjective experiences of good government, while none has explicitly discussed public administration from that context. All of these scholars have made a positive contribution to the endeavor of understanding public administration. However, all have suffered through the problems associated with being as close as they have been without having a complete solution (understanding the process while not being able to encompass the form). It merits the time for us to examine each scholar's contribution briefly, place it in the context of Evolutionary Critical Theory, and demonstrate how it solves, completes, and enhances their work.

Rohr argues that public administration is subordinate to each of the branches of government while being able to choose from its constitutional "masters" from moment to moment.[59] He uses the image of a mechanistic "balance wheel" to demonstrate how this process might happen, but mostly as a communicative device. The important element of Rohr's work comes first from his understanding of the complexity of relationships among the branches of government, public administration, and society, and second from his insight that it is the *process* rather than the *structure* that is essential to public administration. By stating that public administration "chooses" which branch to be responsible to, Rohr argues for the type of situational "power" that creates a space for the experience of governance, and it is this space for governance, not the "wheel" or some other communicative artifact, that legitimates public administration both in theory and practice.

Terry wrestles similarly with the means to articulate the relationship between public administration and society.[60] Over the years, he has attempted to describe, explain, and understand public administration from several approaches, many of which build outward from Rohr's position. When Terry describes a "conservator" as a vessel of public values, he implicitly argues that public administration creates a space for and fosters the experience of good governance that we have articulated throughout this book. In his later work, Terry also begins to apprehend the notions of endogenous evolution and drift, attempting to encapsulate them in a theater metaphor that includes both the positive and negative elements of public service.[61] Terry, then, moves us even a bit closer to our current understanding by demonstrating a need for understanding these intersubjective experiences of good governance.

It appears that Terry remains uncomfortable with some of the subjective aspects of his work. The collaborative work with Spicer tends to illustrate this.[62] During these collaborations, Spicer and Terry wrestle with the incremental nature of what we describe as endogenous evolution through the use of common law reasoning, critique of motives, and other procedural mechanisms. Their efforts have faced challenges, including, for example, the arguments that the Constitution is more than a "blueprint or algorithm for government" obscuring how it confirmed a social order,[63] and that their work "meekly returned to a far less interesting theme of hypocrisy" that questioned the "Lockean contractarian principles of the framers which Spicer and Terry themselves accept."[64] Essentially, Spicer and Terry find themselves under attack because of the teleology of their argument. If they instead were able to address the problems of romanticism, slavery, or incrementalism from a nonteleological (evolutionary) approach, we might then leave many of the artifacts, issues, and debates undermining their arguments where they often belong, in the dust of a prior intersubjective experience.

Epistemology

The final, but by no means the least of the areas we discuss in this chapter is epistemology. Specifically, public administration tends to be divided into roughly two camps, the "empiricists" and the "theorists." We call one group "empiricists" simply because their primary mode of operation is within the realm of empirical (traditional) social scientists. They hold and have internalized the beliefs of what is commonly understood as the functionalist paradigm.[65] In contrast, the "theorists" tend to at least dabble in the interpretive paradigm (primarily through phenomenology and hermeneutics)[66] and, to a lesser extent, in radical structuralism (including contemporary Marxism)[67] or radical humanism.[68] In many cases, there tends to be broad disagreement among empiricists and theorists as we have conceived them here, particularly about what form knowledge should take. In addition, there has not been an adequate discussion of either ontology or epistemology (though we hope to have advanced this somewhat) in public administration outside a narrow group of discussions by a few select authors.[69] Most recently, the primary discussion has been about how "quality" of research often can justify or at least support some of the arguments against public administration's identity apart from political science or business.[70]

This lack of development, then forces us to address several aspects of the epistemology of public administration. By reflecting on our work throughout this book, we have demonstrated that these divisions among empiricists and theorists are at best artificial and at worst counterproductive, given the synergistic nature of Evolutionary Critical Theory. Rather than focusing on the creation of argumentative seams, by using Evolutionary Critical Theory we can understand how both groups can exist simultaneously, since although much of what we do in public administration is constructed, the intersubjective experiences remain very "real."

This statement can be supplemented by realizing that there are different types, different "brands" of empiricism (traditional social science).[71] These brands vary from "eternal objects" to "knowledge based on experience," encompassing a range of relatively clear to relatively vague terms understood as "descriptive" or "suppositional" symbols.[72] Specifically, Benjamin discusses the "named" brands positivism, fictionalism, and realism. In his mind, fictionalism and realism are the two most amenable to a very narrow understanding of "constructionism."[73] The constructionism posited by Benjamin is what we would identify today as the symbolic positivism that relies on indirect measurements for a phenomenon that is *believed* to be real and measurable. This, as a consequence, supports our argument that even within what traditional social science considers real there can be drift toward more constructed, more subjective conceptions.

If we take into account the notion of measurement errors and debates about the principle of uncertainty, including debates about whether or not causality is possible, we can further our argument for synergy among the paradigms. By understanding that we cannot measure both position and momentum simultaneously, and that if we focus on precisely measuring one, we sacrifice understanding more about the other,[74] we then can begin to understand the limits of empirical measurements, even if we assume a state of positivist empiricism. Furthermore, Darwin argues that "knowledge is no use in predicting what is going to happen later," even though he maintains that we "can not say exactly what will happen to a single electron, but we can confidently estimate the probabilities"[75] which is really not confidence-inspiring in even the broadest philosophical terms, given that the notion of indeterminism has not truly been resolved as a philosophical, ontological, or empistemological endeavor among the "positivists."

We then must consider the question of what is the relationship between knowledge and reality? In other words, is there a means to understand the subjective and the objective within a similar context? The problem at the core of such a question is one of teleology, in that there is a separation of "positive theory of knowledge from the metaphysics of knowledge."[76] Eaton arguably made one of the first reasonable efforts to examine knowledge outside of the teleological boundaries created by being a critical realist, neorealist, or idealist, or what he calls a Kantian, Bergsonian, or Hegelian.[77] Eaton makes the case that it might not be appropriate to presuppose "that truth *must* be a revelation of reality," particularly if we consider that knowledge might be a knowledge of appearances, not realities themselves.[78] In addition, he argues that truth is based on agreement,[79] which, arguably, reinforces our argument about the relationships among knowledge, truth, and reality in Evolutionary Critical Theory.

Considering the limits of empiricism, understood as "behaviorism," as "science," or as some other label, we can then unveil a more reasonable understanding of the relationship among these seemingly disparate ideas of functionalism, interpretivism, radical structuralism, and radical humanism.[80] Nelson argues that since there is not a complete agreement on inquiry (on truth), then any such limited description would be too "weak" to describe even the performance of some automaton or other basic observable mechanism.[81] In addition, since the behavioral sciences must work in the context of space and time, then we can realize, even in the most trivial sense of understanding, that empiricism *must, by its nature,* suffer from the effects of measurement proffered by Heisenberg[82] and elucidated by Darwin.[83] However, if we pursue Popper's argument, that human reason allows for the creation of objective standards of criticism, which enables people to *act, contemplate, discriminate,* and *adjudicate,* then we might bridge this chasm among these seemingly disparate beliefs.[84] Popper even alludes to the means for achieving this standard of knowledge: "inter-subjective criticism."[85] By considering intersubjectivity as a means of understanding the world, Popper ultimately gives support to our argument in Chapter 2.

Thus, if we now reflect on Evolutionary Critical Theory and public administration together, we can see that the intersubjective experience of good government provides the justification for how this reconstructed critical theory synergistically bridges the functionalist, interpretivist, structuralist, and humanist approaches to knowledge. By focusing on intersubjectivity, based on human reason and negotiation, we create a shared space both for empiricists (by providing a "real" means for getting at some objective truth, à la Popper), while retaining space for the function of immanent critique, social construction, phenomenology, and other "subjective" approaches that work best when we have reached the "seams" of empirical logic, due to measurement, complexity, incommeasurability, or other phenomena broader than the scope of empirical reasoning.

Concluding Remarks

We may now contend that any number of these arguments, debates, and impasses about identity, legitimacy, and epistemology in public administration represent what is best described as a communication breakdown. Such a communication breakdown grows from our orthodox understanding of teleology, epistemology, and socialization within a common frame. However, if we use critical theory generally, and Evolutionary Critical Theory in particular, we might forge a more complete understanding of public administration as a discipline, as a practice, and as a societal entity. This understanding allows us to build the conceptual bridges that enable Evolutionary Critical Theory to act as the means to create a sort of "solidarity"[86] among these seemingly disparate worldviews or beliefs. We can then effectively deconstruct, recover, and reconstruct these shards of difference within the disciplinary matrix of public administration, crafting a lifeworld within the intersubjective experiences of good governance.

Such a reestablishment or reenvisioning of theory and praxis in public administration within Evolutionary Critical Theory necessitates the renegotiation of any number of the concepts we have discussed in the latter parts of this book. Evolutionary Critical Theory is not a means to patch together a plethora of half-theories, beliefs, and assertions. It instead requires continual deconstruction, reflection, and reconstruction to create meaning, emancipation, and a good society through the process of endogenous evolution *within* an institutional context, rather than from its collapse. This, in turn, makes public administration the place of possibilities instead of a place of nihilism or despair.

Public administration's future, then, is wide open. What we choose to make it will be a function of our experiences, understandings, beliefs, and ideals. As we translate them into theory and praxis, we then become the custodians of societal knowledge, the caretakers of power, emancipation, and the shared space to address human concerns.

Notes

Notes to Chapter 1

1. L. Lynn, C. Heinrich, and C. Hilt, "The Empirical Study of Governance: Theories, Models, Methods" (presented at the Workshop on Models and Methods for the Empirical Study of Governance, University of Arizona, Tucson, April 29–May 1, 1999), p. 1.
2. J. Gunnell, "Why There Cannot Be a Theory of Politics," *Polity* 29 (4): 531.
3. Lynn, Heinrich, and Hilt, "Empirical Study of Governance," p. 4.
4. B. Peters, and J. Pierre, "Governance Without Government? Rethinking Public Administration," *Journal of Public Administration Research and Theory* 8 (2): 223–43.
5. Lynn, Heinrich, and Hilt, "Empirical Study of Governance," p. 3.
6. C. Wilber, and R. Harrison, "The Methodological Basis of Institutional Economics: Pattern Model, Storytelling, and Holism," *Journal of Economic Issues* 12 (1): 75.
7. Ibid.
8. J. Wisman, and J. Rozansky, "The Methodology of Institutionalism Revisited," *Journal of Economic Issues* 25 (3): 713.
9. W. James, *Principles of Psychology*, vol. 1 (New York: Dorer, 1955), p. 488.
10. G. Heather, and M. Stolz, "Hannah Arendt and the Problem of Critical Theory," *Journal of Politics* 41 (1): 3.
11. D. Kellner, "Critical Theory Today: Revisiting the Classics"; available at www.uta.edu/huma/illuminations/kell10.htm (December 1, 2002).
12. S. Buck-Morss, *The Origin of Negative Dialectics: Theodor W. Adorno, Walter Benjamin and the Frankfurt Institute* (New York: Free Press, 1977); S. Breuer, "Adorno's Anthropology," *Teleos* 64 (Summer): 15–32.
13. J. Habermas, *Knowledge and Human Interests* (Cambridge, UK: Polity Press, 1987); T.W. Adorno, *Negative Dialectics* (New York: Seabury, 1973); T. Adorno, "Introduction," in *The Positivist Dispute in German Sociology* (London: Heinemann, 1979); H. Marcuse, *One Dimensional Man: Studies in the Ideology of Advanced Industrial Society* (Boston: Beacon Press, 1964); E. Fromm, "Character and the Social Process," in *Fear of Freedom* (New York: Routledge, 1942); D. Shalin, "Critical Theory and the Pragmatist Challenge," *American Journal of Sociology* 98 (2): 237–79; A. Honneth, "Integrity and Disrespect: Principles of a Conception of Morality Based on a Theory of Recognition," in *The Fragmented World of the Social* (Albany: State University of New York Press, 1995); J. Dryzek, "Discursive Designs: Critical Theory and Political Institutions," *American Journal of Political Science* 31 (3): 662–63; A.

Honneth, "The Social Dynamics of Disrespect: On the Location of Critical Theory Today," trans. John Farrell, *Constellations* 1 (2): 255–69; T. Schroyer, *The Critique of Domination* (New York: Braziller, 1973), pp. 132–68.

14. M. Horkheimer, and T. Adorno, *Dialectic of Enlightenment* (New York: Herder and Herder, 1972), p. xi.

15. J. March, and H. Simon, *Organizations* (New York: John Wiley, 1958); A. Wildavsky, *Speaking Truth to Power* (Boston: Little, Brown, 1979); M. Landau, "Redundancy, Rationality and the Problem of Duplication and Overlap," *Public Administration Review* 19: 79–88; H. Kaufman, "Administrative Decentralization and Political Power," *Public Administration Review* 29: 3–15; C. Gould, "Private Rights and Public Virtues: Women, the Family, and Democracy," in *Beyond Domination: New Perspectives on Women and Philosophy*, ed, C.G. Gould (Totowa, NJ: Rowman & Allanheld, 1984); J. Hearn, *The Gender of Oppression: Men, Masculinity, and the Critique of Marxism* (Brighton, UK: Wheatsheaf Books, 1987); B. Adam, *The Survival of Domination: Inferiorization and Everyday Life* (New York: Elsevier North-Holland, 1978); A. Memmi, *Dominated Man: Notes Towards a Portrait* (Boston: Beacon, 1968); C. Abel, and F. Marsh, *Political Trials: Criticisms and Justifications* (Westport, CT: Greenwood Press, 1993).

16. D. Kellner, "Critical Theory Today"; J. Schumpeter, *Capitalism, Socialism and Democracy*, 3d ed. (New York: Harper and Row, 1950), pp. 235–68; Z. Tar, *The Frankfurt School: The Critical Theories of Max Horkheimer and Theodor W. Adorno* (New York: Wiley, 1977); M. Jay, *The Dialectical Imagination: A History of the Frankfurt School and the Institute of Social Research 1923–1950* (Toronto: Little, Brown, 1973).

17. A. Honneth, and J. Farrell, "Recognition and Moral Obligation," *Social Research* 64 (1): 7.

18. J. Habermas, "On the Pragmatic, the Ethical, and the Moral Employments of Practical Reason," in *Habermas, Justification and Application* (Boston: MIT Press, 1995), pp. 1–18.

19. Tar, *Frankfurt School*; Jay, *Dialectical Imagination*.

20. R. Golembiewski, W. Welsch, and W. Crotty, W., *A Methodological Primer for Political Scientists* (Chicago: Rand McNally, 1969); P. Kronenberg, "The Scientific and Moral Authority of Empirical Theory of Public Administration," in *Toward a New Public Administration: The Minnowbrook Perspective*, ed. F. Marini (Scranton, PA: Chandler, 1971); V. Ostrom, *The Intellectual Crisis in American Public Administration*, 2d ed. (Tuscaloosa: University of Alabama Press, 1989); J.L. Perry, and K.L. Kraemer, "Research Methodology in the Public Administration: Issues and Patterns, in *Public Administration: The State of the Discipline* (Chatham, NJ: Chatham House, 1990), pp. 347–72; M. Whicker, R. Strickland, and D. Olshfski, "The Troublesome Cleft: Public Administration and Political Science," *Public Administration Review* 53 (6): 531–41.

21. E. Albaek, "Between Knowledge and Power: Utilization of Social Science in Public Policymaking," *Policy Sciences* 28 (1): 79–100; P.L. Berger, and T. Luckman, *The Social Construction of Reality* (New York: Doubleday, 1966); T. Bhaskar, *A Realist Theory of Science* (Leeds, UK: Leeds Books, 1975); M. Danziger, "Policy Analysis Postmodernized: Some Political and Pedagogical Ramifications," *Policy Studies Journal* 23 (3): 435–50; J. Habermas, *Between Facts and Norm: Contributions to a Discourse Theory of Law and Democracy* (Cambridge, MA: MIT Press, 1996); M. Harmon, *Responsibility as Paradox: A Critique of Rational*

Discourse on Government (Thousand Oaks, CA: Sage, 1995); A. Lin, "Bridging Positivist and Interpretivist Approaches to Qualitative Methods," *Policy Studies Journal* 26 (1): 162–80; G. Marshall, "Deconstructing Administrative Behavior: The "Real" as Representation, *Administrative Theory and Praxis* 18: 117–27; O.C. McSwite, *Legitimacy in Public Administration: A Discourse Analysis* (Thousand Oaks, CA: Sage, 1997); G. Morçöl, "A Meno Paradox for Public Administration: Have We Acquired a Radically New Knowledge from the 'New Sciences'?" *Administrative Theory and Praxis* 19 (3): 305–17; D. Torgerson, "Between Knowledge and Politics," *Policy Sciences* 19: 33–59; G. Wamsley, and J. Wolf, *Refounding Democratic Public Administration: Modern Paradoxes, Postmodern Challenges* (London: Sage, 1996).

22. L. Keller, and M. Spicer, "Political Science and American Public Administration: A Necessary Cleft?" *Public Administration Review* 57 (3): 270; L. Mainzer, "Public Administration in Search of a Theory: The Interdisciplinary Delusion," *Administration and Society* 26 (3): 359–94; Whicker, Strickland, and Olshfski, "Troublesome Cleft."

Notes to Chapter 2

1. A. Donagan, "Historical Explanation: The Popper-Hempel Theory Reconsidered," *History and Theory* 6: 3–26; J. Nelson, "Accidents, Laws, and Philosophic Flaws: Behavioral Explanations in Dahl and Dahrendorf," *Comparative Politics* 7: 435–57.

2. S. Verba, "Comparative Politics: Where Have We Been, Where Are We Going?" In *New Directions in Comparative Politics*, ed. H.J. Wiarda (Boulder, CO: Westview, 1985), p. 34.

3. D. Waldo, "Public Administration." *Journal of Politics* 30 (2): 459.

4. R. Merton, *Social Theory and Social Structure* (New York: Free Press, 1957), pp. 5–6.

5. R. Beiner, "Review of Peter J. Steinberger's *The Concept of Political Judgment*," *Political Theory* 22 (November): 688.

6. R. Beiner, *Political Judgment* (Chicago: University of Chicago Press, 1983), p. xv.

7. F. Marini, *Toward a New Public Administration: The Minnowbrook Perspective* (Scranton, PA: Chandler, 1971); R. Box, "An Examination of the Debate Over Research in Public Administration," *Public Administration Review* 52: 62–69; J. White, G. Adams, and J. Forrester, "Knowledge and Theory Development in Public Administration: The Role of Doctoral Education and Research," *Public Administration Review* 56: 441–52.

8. M. Rutgers, "Paradigm Lost: Crisis as Identity of the Study of Public Administration," *International Review of Administrative Sciences* 64 (4): 553–64.

9. Dahl, R., "The Science of Public Administration: Three Problems" *Public Administration Review* 7 (Winter): 1–11.

10. H.A. Simon, "The Proverbs of Administration," *Public Administration Review* 6: 53–67.

11. D. Waldo, "Scope of the Theory of Public Administration," in *Theory and Practice of Public Administration: Scope, Objectives, and Methods*, ed. James C. Charlesworth (Philadelphia: American Academy of Political and Social Science, 1968), pp. 1–26; Waldo, "Public Administration," pp. 443–79; D. Waldo, *The Administrative*

State: A Study of the Political Theory of American Public Administration (New York: Holmes and Meier, 1984).

12. T. Kuhn, *The Structure of Scientific Revolutions*, 2d ed. (Chicago: University of Chicago Press, 1970), p. 182.

13. Ibid., pp. 186–91.

14. R. Montjoy, and D. Watson, "A Case for Reinterpreted Dichotomy of Politics and Administration as a Professional Standard in Council-Manager Government," *Public Administration Review* 55 (3): 231–39.

15. G. Garvey, "False Promises: The NPR in Historical Perspective," in *Inside the Reinvention Machine: Appraising Governmental Reform*, ed. D. Kettl and J. DiIulio (Washington, DC: Brookings Institution, 1995), p. 87.

16. Waldo, "Scope of the Theory of Public Administration."

17. Waldo, *Administrative State*.

18. V. Ostrom, *The Intellectual Crisis in American Public Administration*, 2d ed. (Tuscaloosa: University of Alabama Press, 1989), pp. 14–18.

19. R. Golembiewski, *Public Administration as a Developing Discipline: Part 1, Perspectives on Past and Present* (New York: Marcel Dekker, 1977).

20. K. Henderson, *Emerging Synthesis in American Public Administration* (New York: Asia Publishing House, 1966).

21. D. Rosenbloom, *Public Administration* (New York: Random House, 1990).

22. D. Farmer, "Derrida, Deconstruction, and Public Administration." *American Behavioral Scientist* 31 (1): 16–17.

23. Ibid., p. 17.

24. Ibid., p. 22.

25. J. Perry, and K. Kraemer, "Research Methodology, in the *Public Administration Review*, 1975–1984," *Public Administration Review* 46 (3): 221.

26. J. White, "Knowledge Development and Use in Public Administration: View from Postpositivism, Poststructuralism, and Postmodernism," in *Public Management in an Interconnected World*, ed. M.T. Bailey and R.T. Mayer, (New York: Greenwood, 1992), p. 175.

27. L. Wittgenstein, *Philosophical Investigations* (New York: Macmillan, 1953).

28. Ibid., p. 31.

29. Ibid., p. 32.

30. L. Wittgenstein, *Zettel* (Oxford: Oxford University Press, 1981), p. 173.

31. Wittgenstein, *Philosophical Investigations*, p. 23.

32. L. Wittgenstein, *Philosophical Grammar* (Oxford: Oxford University Press, 1974), p. 59.

33. Wittgenstein, *Philosophical Investigations*, note 47, p. 67.

34. D. Farmer, *The Language of Public Administration: Bureaucracy, Modernity and Postmodernity* (Tuscaloosa: University of Alabama Press, 1995).

35. Ibid., p. 4.

36. D. Farmer, "Public Administration Discourse: A Matter of Style?" *Administration and Society* 31 (3): 313.

37. Farmer, *Language of Public Administration*, p. 13.

38. C. Fox, and H. Miller, "The Epistemic Community," *Administration and Society* 32 (6): 668.

39. Ibid.

40. Ibid., p. 676.

41. Ibid., p. 678.

42. Ibid., p. 679.
43. K. Hansen, "Identifying Facets of Democratic Administration," *Administration and Society* 30 (4): 444–45.
44. Fox and Miller, "Epistemic Community."
45. K. Hansen, "Clarifying the Assumptions: The Empirical Referents of Discourse Revisited," *Administration and Society* 30 (5): 604.
46. Wittgenstein, *Philosophical Grammar*, p. 59.
47. Kuhn, *Structure of Scientific Revolutions*.
48. M. Rutgers, "Beyond Woodrow Wilson: The Identity of the Study of Public Administration in Historical Perspective," *Administration and Society* 29 (3): 290.
49. Ibid., p. 291.
50. L. White, *Introduction to the Study of Public Administration* (New York: Macmillan, 1926), p. 2.
51. Waldo, *Administrative State*, p. 448.
52. W. Wilson, "The Study of Administration," in *The Papers of Woodrow Wilson*, vol. 5 (Princeton, NJ: Princeton University Press, 1966), p. 372.
53. R. Miewald, "The Origins of Wilson's Thought: The German Tradition and the Organic State," in *Politics and Administration*, ed. J. Rabin and J.S. Bowman (New York: Marcel Dekker, 1984), pp. 17–30.
54. Wilson, "Study of Administration," pp. 371–73.
55. Waldo, *Administrative State*, p. 108.
56. R. Golembiewski, *Public Administration as a Developing Discipline*, p. 9.
57. G. Caiden, "In Search of an Apolitical Science of American Public Administration," in ed. J. Rabin and J.S. Bowman, *Politics and Administration* (New York: Marcel Dekker, 1984), p. 60.
58. F. Goodnow, *Politics and Administration* (New York: Macmillan, 1900), p. 73.
59. Wilson, "Study of Administration," p. 17.
60. Goodnow, *Politics and Administration*, pp. 113–14.
61. Wilson, "Study of Administration," p. 23.
62. Waldo, *Administrative State*.
63. W. Wilson, "The Study of Public Administration," *Political Science Quarterly* 2 (June): 197–222.
64. P. Van Riper, "The Politics-Administration Dichotomy: Concept or Reality?" in *Politics and Administration*, ed. J. Rabin and J.S. Bowman (New York: Marcel Dekker, 1984), pp. 203–18.
65. Miewald, "Origins of Wilson's Thought, pp. 25–26.
66. R. Stillman, "Woodrow Wilson and the Study of Administration," *American Political Science Review* 67(6): 582–88.
67. Farmer, "Public Administration Discourse: A Matter of Style?" p. 324.
68. J. Rabin, and J. Bowman, eds., *Politics and Administration: Woodrow Wilson and American Public Administration* (New York: Marcel Dekker, 1984), p. 4.
69. Waldo, *Administrative State*, p. 128.
70. P. Appleby, *Policy and Administration* (University: University of Alabama Press, 1949), p. 26.
71. N. Long, "Power and Administration," *Public Administration Review* 9 (4): 257.
72. L. Gulick, and L. Urwick, eds., *Papers on the Science of Administration* (New York: Institute of Public Administration, 1937), pp. 10, 192.
73. Cited in Waldo, "Public Administration," pp. 443–79.

74. R. Dahl, "The Science of Public Administration: Three Problems," *Public Administration Review* 7 (Winter): 1–11.

75. Ostrom, *The Intellectual Crisis in American Public Administration*.

76. Ibid., pp. 50–51.

77. R. Golembiewski, "A Critique of 'Democratic Administration' and its Supporting Ideation," *American Political Science Review* 71 (4): 1488–1507.

78. G. Hardin, "The Tragedy of the Commons," *Science* 162: 1243–48; M. Olson, *The Logic of Collective Action: Public Goods and the Theory of Groups* (Cambridge, MA: Harvard University Press, 1965).

79. E. Wald, and I. Hoos, "Toward a Paradigm of Future Public Administration," *Public Administration Review* 33 (4): 366.

80. Ibid., p. 372.

81. R. Bingham, and W. Bowen, "'Mainstream' Public Administration Over Time: A Topical Content Analysis of *Public Administration Review*," *Public Administration Review* 54 (2): 204–8.

82. Z. Lan, and K. Anders, "A Paradigmatic View of Contemporary Public Administration Research: An Empirical Test," *Administration and Society* 32 (2): 138–65.

83. Ibid., p. 142.

84. Ibid., p. 151.

85. Ibid., p. 153.

86. Ibid., p. 154.

87. Kuhn, *Structure of Scientific Revolutions*, p. 182.

88. Ibid., pp. 186–91.

89. Lan and Anders, "A Paradigmatic View of Contemporary Public Administration Research," p. 165.

90. Ibid., p. 155.

Notes to Chapter 3

1. P. Winch, *The Idea of a Social Science and Its Relationship to Philosophy* (London: Routledge and Kegan Paul, 1958), p. 73; C. Taylor, "Interpretation and the Sciences of Man," in *Critical Sociology*, ed. P. Connerton (Harmondsworth, UK: Penguin, 1976).

2. See, for example, J. Gunnell, "Realizing Theory: The Philosophy of Science Revisited," *Journal of Politics* 57 (4): 923–40.

3. J. Gunnell, "Why There Cannot Be a Theory of Politics," *Polity* 29 (4): 531.

4. B. Russell, *History of Western Philosophy* (New York: Simon and Schuster, 1945), p. 33.

5. P. Bill, "Notes on the Greek Theoros and Theoria," *Transactions of the American Philological Association* 32: 196; K. Kerényi, *The Religion of the Greeks and Romans* (Stockholm: Thames & Hudson, 1962).

6. Plato, *The Republic* (New York: Oxford University Press, 1945).

7. K. Deutsch, "On Political Theory and Political Action," *American Political Science Review* 65 (1): 11.

8. Ibid., p. 12.

9. Ibid.

10. C. Taylor, *Sources of the Self: The Making of the Modern Identity* (Cambridge, MA: Harvard University Press, 1989), pp. 3–4, 18.

11. L. Wittgenstein, *Philosophical Investigations* (Oxford: Oxford University Press, 1953).
12. Gunnell, "Why There Cannot Be a Theory of Politics," p. 529.
13. Ibid., p. 530.
14. C. Wilber, and R. Harrison, "The Methodological Basis of Institutional Economics: Pattern Model, Storytelling, and Holism," *Journal of Economic Issues* 12 (1): 75.
15. J. Wisman, and J. Rozansky, "The Methodology of Institutionalism Revisited," *Journal of Economic Issues* 25 (3): 713.
16. R. Merton, *Social Theory and Social Structure*, rev. ed. (New York: Free Press of Glencoe, 1957), 5–6.
17. J. Heap, and P. Roth, "On Phenomenological Sociology," *American Sociological Review* 38 (3): 354–67.
18. See, for example, W. James, *The Varieties of Religious Experience* (New York: Touchstone Books, 1997).
19. D. Ihde, "Introduction," in P. Ricoeur's, *Hermeneutics and the Human Sciences*, ed. J.B. Thompson (Cambridge University Press, 1981), p. xvii.
20. H. Garfinkel, *Studies in Ethnomethodology* (Englewood Cliffs, NJ: Prentice-Hall, 1967).
21. A. Schutz, *The Phenomenology of the Social World* (Evanston, IL: Northwestern University Press, 1967), p. 11.
22. Ibid., p. 14.
23. E. Nagel, *The Structure of Science* (New York: Harcourt, Brace and World, 1961).
24. M. Heidegger, *Being and Time* (New York: Harper and Row, 1962), p. 155.
25. E. Husserl, *The Ideas* (New York: Collier, 1962), sections, 30–32.
26. R. Kearney, *Modern Movements in European Philosophy*, 2d ed. (Manchester: Manchester University Press, 1994), p. 16.
27. J. Caputo, "The Question of Being and Transcendental Phenomenology: Reflections on Heidegger's Relationship to Husserl," in *Radical Phenomenology: Essays in Honor of Martin Heidegger*, ed. J. Sallis (Atlantic Highlands, NJ: Humanities, 1978), note 5, p. 86.
28. M. Warnock, *Existentialism* (Oxford: Oxford University Press, 1970), p. 35.
29. A. Schutz, "The Problem of Transcendental Intersubjectivity in Husserl," *Collected Papers*, vol. 3, *Studies in Phenomenological Philosophy*, ed. I. Schutz (The Hague: Nijhoff, 1966). (Italics added.)
30. Ibid.
31. Ibid., p. 82.
32. E. Husserl, *The Idea of Phenomenology* (The Hague: Nijhoff, 1970), pp. 59–60.
33. E. Husserl, *The Crisis of European Sciences and Transcendental Phenomenology* (Evanston, IL: Northwestern University Press, 1970), pp. 138–39.
34. Husserl, *Idea of Phenomenology*, pp. 7–9.
35. E. Durkheim, *The Elementary Forms of the Religious Life* (London: George Allen and Unwin, 1915).
36. Garfinkel, *Studies in Ethnomethodology*.
37. Ibid.
38. Durkheim, *Elementary Forms of the Religious Life*.
39. P. Berger, and T. Luckmann, *The Social Construction of Reality: A Treatise in the Sociology of Knowledge* (London: Penguin, 1966).

40. T. Adorno, *Negative Dialectics* (New York: Continuum, 1973).
41. M. Horkheimer, and T. Adorno, *The Dialectics of Enlightenment* (London: Verso, 1947).
42. J. Habermas, *The Theory of Communicative Action* (Boston: Beacon, 1984); Horkheimer, and Adorno, *Dialectics of Enlightenment*; B. Fay, *Critical Social Science* (Cambridge, UK: Polity, 1987); E. Fromm, *To Have or To Be?* (New York: Harper and Brothers, 1976).
43. B. Wheeler, "Antisemitism as Distorted Politics: Adorno on the Public Sphere," *Jewish Social Studies* 7 (2): 114.
44. Ibid.
45. Ibid.
46. Habermas, *Theory of Communicative Action*; J. Habermas, *Knowledge and Human Interests* (Cambridge, UK: Polity, 1987).
47. E. Fields, "Understanding Activist Fundamentalism: Capitalist Crisis and the Colonization of the Lifeworld," *Sociological Analysis* 52: 175.
48. Wheeler, "Antisemitism as Distorted Politics," *Jewish Social Studies* 7 (2): 134.
49. S. Bruce, "Fundamentalism, Ethnicity and Enclave," in *Fundamentalisms and the State: Remaking Polities, Economies, and Militance*, ed. M. Marty and R.S. Appleby (Chicago: University of Chicago Press, 1993), pp. 50–67.
50. W. Connolly, "Beyond Good and Evil: The Ethical Sensibility of Michel Foucault," *Political Theory* 21: 365–89.
51. W. Connolly, *The Ethos of Pluralization* (Minneapolis: University of Minnesota Press, 1995), p. 93.
52. H. Simon, "Why Public Administration?" *Journal of Public Administration Research and Theory* 8 (1): p. 6.
53. J. Locke, *Two Treatises of Government* (New York: Dutton, 1978); R. Harré, and E. Madden, *Causal Powers: A Theory of Natural Necessity* (Oxford: Basil Blackwell, 1975).
54. S. Shoemaker, "Causality and Properties," in *Time and Cause*, ed. Peter Van Inwagen (Dordrecht: D. Reidel, 1980), pp. 109–36.
55. C. Macpherson, *Democratic Theory: Essays in Retrieval* (Oxford: Oxford University Press, 1973).
56. T. Newton, "Theorizing Subjectivity in Organizations: The Failure of Foucauldian Studies?" *Organization Studies* 19 (3): 415–47.
57. J. Searle, *Minds, Brains, and Science* (Cambridge, MA: Harvard University Press, 1984), p. 94.
58. T. Reid, *Essays on the Active Powers of the Human Mind* (Cambridge, MA: MIT Press, 1969), p. 367.
59. Heidegger, *Being and Time*, p. 155.
60. M. Foucault, "The Ethics of Care for the Self as a Practice of Freedom," *Philosophy and Social Criticism* 12: 122.
61. H. Plessner, "With Different Eyes," in *Phenomenology and Sociology*, ed. Thomas Luckmann (New York: Penguin, 1978), p. 39.
62. P. Ricoeur, *Hermeneutics and the Human Sciences,* trans. J.B. Thompson, (Cambridge: Cambridge University Press, 1981), p. 17.
63. R. Rorty, *Contingency, Irony, and Solidarity* (New York: Cambridge University Press, 1989); B. Smith, *Contingencies of Value: Alternative Perspectives for Critical Theory* (Cambridge, MA: Harvard University Press, 1988).

64. I. Prigogine, and I. Stengers, *Order Out of Chaos: Man's New Dialogue with Nature* (New York: Bantam Books, 1984), p. 313.
65. Ibid.
66. Ibid.
67. Ibid.
68. M. Scriven, "Logical Positivism and the Behavioral Sciences," in *The Legacy of Logical Positivism: Studies in the Philosophy of Science*, ed. P. Achinstein and S. Barker (Baltimore, MD: Johns Hopkins University Press, 1969).
69. L. Wittgenstein, *On Certainty* (New York: Harper Torchbooks, 1969), p. 81.
70. Ibid.
71. Gunnell, "Why There Cannot Be a Theory of Politics," p. 528.
72. A. Hamilton, J. Madison, and J. Jay, *The Federalist Papers* (New York: Bantam Books, 1982).
73. R. Beiner, *What's the Matter with Liberalism?* (Berkeley: University of California Press, 1992), p. 108.
74. M. Glendon, *Rights Talk* (New York: Free Press, 1991); M. Walzer, "Democracy and Philosophy," *Political Theory* 9: 379–99.
75. Connolly, *Ethos of Pluralization*, p. 103.
76. C. King, K. Felty, and B. Susel, "The Questions of Participation: Toward Authentic Public Participation in Public Administration," *Public Administration Review* 58 (4): 317–26; C. King, "Talking Beyond the Rational," *American Review Of Public Administration* 30 (3): 271–91; R. Hummel, "Critique of 'Public Space,'" *Administration and Society* 34 (1): 102–7; O. Mcswite, *Legitimacy in Public Administration: A Discourse Analysis* (Thousand Oaks, CA: Sage, 1997); C. Stivers, "Toward Administrative Public Space: Hannah Arendt Meets the Municipal Housekeepers," *Administration and Society* 34 (1): 98–101; G. Veenstra, "Explicating Social Capital: Trust and Participation in Civil Space," *Canadian Journal of Sociology* 27 (4): 547–72.
77. J. Shklar, *The Faces of Injustice* (New Haven: Yale University Press, 1990), p. 83; R. Solomon, *A Passion for Justice* (New York: Addison-Wesley, 1990); F. Waal, *The Origins of Right and Wrong in Humans and Other Animals* (Cambridge, MA: Harvard University Press, 1996).
78. E. Wohlgast, *The Grammar of Justice* (Ithaca, NY: Cornell University Press, 1987), pp. 46–47.
79. Shklar, *Faces of Injustice*; Solomon, *Passion for Justice*.
80. J. Locke, *The Second Treatise on Civil Government* (Amherst, NY: Prometheus, 1986); J. Rousseau, *The Social Contract* (New York: Penguin, 1968).
81. T. Hobbes, *Leviathan* (New York: Penguin, 1968).
82. Locke, *Second Treatise on Civil Government*.
83. Rousseau, *Social Contract*; E. Kant, *The Metaphysics of Morals* (Cambridge, MA: Cambridge University Press, 1996).
84. F. Hegel, *The Philosophy of Right* (Oxford: Oxford University Press, 1967).
85. A. Sementelli, and C.F. Abel, "Recasting Critical Theory: Veblen, Deconstruction and the Theory Practice Gap," *Administrative Theory and Praxis* 22 (3): 458–78.

Notes to Chapter 4

1. B. Wheeler, "Antisemitism as Distorted Politics: Adorno on the Public Sphere," *Jewish Social Studies* 7 (2): 134.

2. J. Street, "The Institutionalist Theory of Economic Development," *Journal of Economic Issues* 21 (December): 1861.

3. J. Wisman, and J. Rozansky, "The Methodology of Institutionalism Revisited," *Journal of Economic Issues* 25 (3): 710, 712, 721.

4. Wisman and Rozansky, "The Methodology of Institutionalism Revisited," p. 719.

5. T. Kisiel, "The Happening of Tradition: The Hermeneutics of Gadamer and Heidegger," in *Hermeneutics and Praxis*, ed. R. Hollinger (Notre Dame, IN: University of Notre Dame Press, 1985; H. Gadamer, *Truth and Method*, trans. J. Weinsheimer and D.G. Marshall (New York: Continuum, 1994); H. Gadamer, *Philosophical Hermeneutics*, trans. D.E. Linge (Berkeley: University of California Press, 1976).

6. D. Farmer, "Public Administration Discourse: A Matter of Style?" *Administration and Society* 31 (3): 324; R. Bingham, and W. Bowen, "'Mainstream' Public Administration Over Time: A Topical Content Analysis of *Public Administration Review*," *Public Administration Review* 54 (2): 204–8.

7. Z. Lan, and K. Anders, "A Paradigmatic View of Contemporary Public Administration Research: An Empirical Test," *Administration and Society* 32 (2): 138–65.

8. R. Beiner, *Political Judgment* (Chicago: University of Chicago Press, 1983), pp. 75–79; B. Barber, *The Conquest of Politics: Liberal Philosophy in Democratic Times* (Princeton, NJ: Princeton University Press, 1988), pp. 151, 205; M. Oakeshott, *Experience and Its Modes* (Cambridge, MA: Cambridge University Press, 1985), pp. 320–21; W. Walzer, "Democracy and Philosophy," *Political Theory* 9 (August): 393.

9. Beiner, *Political Judgment*, p. 139; Barber, *Conquest of Politics*, p. 200.

10. M. Walzer, *Interpretation and Social Criticism* (Cambridge, MA: Harvard University Press, 1987), p. 40.

11. C. Wilber, and R. Harrison, "The Methodological Basis of Institutional Economics: Pattern Model, Storytelling, and Holism," *Journal of Economic Issues* 12 (1): 61–89.

12. Wisman and Rozansky, "Methodology of Institutionalism Revisited," p. 723.

13. Ibid.

14. P. Bush, "The Theory of Institutional Change," *Journal of Economic Issues* 21 (September): 1077.

15. Ibid., p. 1076.

16. Wisman and Rozansky, "Methodology of Institutionalism Revisited," p. 717.

17. Bush, "Theory of Institutional Change," p. 1102.

18. Ibid., pp. 1102–3.

19. T. Veblen, *The Place of Science in Modern Civilization and Other Essays* (New York: Russell and Russell, 1961), p. 239.

20. Wilber and Harrison, "Methodological Basis of Institutional Economics," p. 71.

21. S. Woodbury, "Power in the Labor Market: Institutionalist Approaches to Labor Problems," *Journal of Economic Issues* 21 (December): 1781.

22. L. Wittgenstein, *Philosophical Investigations* (New York: Macmillan, 1953).

23. J. Gunnell, "Why There Cannot Be a Theory of Politics," *Polity* 29 (4): 529.

24. Woodbury, "Power in the Labor Market," pp. 1781–1807.

25. Wisman and Rozansky, "Methodology of Institutionalism Revisited," p. 723.

26. A. Gruchy, *Modern Economic Thought: The American Contribution* (New York: Prentice-Hall, 1947), p. 21.

27. Wisman and Rozansky, "Methodology of Institutionalism Revisited," p. 719.

28. M. Tool, "Introduction," *Journal of Economic Issues* 21 (September): 957.
29. Wisman and Rozansky, "Methodology of Institutionalism Revisited," p. 715.
30. C. Ayres, *The Industrial Economy* (Boston: Houghton Mifflin, 1952), pp. 13, 12, 11, as cited in Wisman and Rozansky, "Methodology of Institutionalism Revisited," p. 716.
31. D. Hamilton, *Evolutionary Economics* (Albuquerque: University of New Mexico Press, 1970), p. 54.
32. S. Hickerson, "Instrumental Valuation: The Normative Compass of Institutional Economics," *Journal of Economic Issues* 21 (September): 1131.
33. Wisman and Rozansky, "Methodology of Institutionalism Revisited," p. 717.
34. P. Reason, and J. Rowan, "Issues of Validity in New Paradigm Research, in *Human Inquiry: A Sourcebook of New Paradigm Research*, ed. P. Reason and J. Rowan (New York: John Wiley, 1981), pp. 239–52.
35. Kisiel, "Happening of Tradition."
36. J. Franklin, "Natural Sciences as Textual Interpretation: The Hermeneutics of the Natural Sign," *Philosophy and Phenomenological Research* 44 (4): 511.
37. Ibid., p. 509.
38. Gadamer, *Truth and Method*; P. Ricoeur, *Hermeneutics and the Human Sciences* (London: Cambridge University Press, 1981).
39. Ibid.
40. Franklin, "Natural Sciences as Textual Interpretation," p. 514.
41. Ibid., p. 517.
42. Wittgenstein, *Philosophical Investigations*.
43. B. Williams, *Ethics and the Limits of Philosophy* (Cambridge, MA: Harvard University Press, 1985).
44. Franklin, "Natural Sciences as Textual Interpretation," p. 517.
45. Woodbury, "Power in the Labor Market," p. 1781.
46. Kisiel, "Happening of Tradition"; Gadamer, *Truth and Method*; Gadamer, *Philosophical Hermeneutics*.
47. Gadamer, *Truth and Method*, p. 269.
48. Ibid., p. 295.
49. H. Wang, *A Logical Journey: From Goedel to Philosophy* (Cambridge, MA: MIT Press, 1997).
50. Wittgenstein, *Philosophical Investigations*, p. 32.
51. L. Wittgenstein, *Zettel* (Oxford: Oxford University Press, 1981), p. 173.
52. L. Wittgenstein, *Philosophical Grammar* (Oxford: Oxford University Press, 1974), p. 59.
53. Wittgenstein, *Philosophical Investigations*, note 47, p. 67.
54. Walzer, *Interpretation and Social Criticism*, p. 40.
55. D. Wagner, and J. Berger, "Do Sociological Theories Grow?" *American Journal of Sociology* 90 (4): 697–728.
56. Ibid., p. 700.
57. Ibid., pp. 697–728.
58. M. Horkheimer, "Traditional and Critical Theory" and "Postscript" [1937], in *Critical Theory: Selected Essays*, trans. Matthew J. O'Connell, (New York: Seabury Press, 1972), pp. 245–46, 215, 252.
59. M. Crozier, "The Frankfurt School," in *Social Theory*, ed. P. Beilharz (North Sydney, Australia: Allen and Unwin, 1991), pp. 90–98; D. Held, *Introduction to Critical Theory: Horkheimer to Habermas* (Berkeley: University of California, 1980).

60. Wittgenstein, *Philosophical Investigations*.
61. Horkheimer, "Traditional and Critical Theory."
62. Horkheimer, "Traditional and Critical Theory" and "Postscript," pp. 245–46, 215, 252.
63. A. Honneth, "The Social Dynamics of Disrespect: On the Location of Critical Theory Today" *Constellations* 1: 255–69.
64. M. Jay, *The Dialectical Imagination* (Boston: Little, Brown, 1973), chapter 2.
65. L. Dennard, "The Democratic Potential in the Transition of Postmodernism," *American Behavioral Scientist* 41 (1): 148–62.
66. C. King, "Healing the Scholarship/Practice Wounds," *Administrative Theory and Praxis* 20 (2): 159–71; Dennard, "Democratic Potential in the Transition of Postmodernism," pp. 148–62.
67. A. Wylie, "Introduction: Socio-Political Context," in *Critical Traditions in Contemporary Archaeology*, ed. V. Pinsky and A. Wylie (Cambridge, MA: Cambridge University Press, 1989), pp. 93–95.
68. A. Schneider, and H. Ingram, "Social Construction of Target Populations: Implications for Politics and Policy," *American Political Science Review* 87 (2): 335.
69. B. Barnes, *Interests and the Growth of Knowledge* (London: Routledge; B. Barnes, *About Science* (Oxford: Basil Blackwell, 1985).
70. T. Adorno, *Negative Dialectics* (London: Routledge, 1973), p. 147.
71. H. Marcuse, *Reason and Revolution* (Boston: Beacon, 1941).
72. Horkheimer, "Traditional and Critical Theory."
73. H. Marcuse, *One Dimensional Man* (Boston: Beacon, 1964), pp. 170–203.
74. D. Cook, "Adorno, Ideology and Ideology Critique," *Philosophy and Social Criticism* 27 (1): 7–8.
75. T. Adorno, *Prisms* (London: Neville Spearman, 1978).
76. M. Horkheimer, *Eclipse of Reason* (New York: Seabury, 1974), p. 182.
77. Held, *Introduction to Critical Theory*, p. 184.
78. T. Adorno, "Cultural Criticism and Society," in *Prisms* (Cambridge, MA: MIT Press, 1967), p. 32.
79. J. Habermas, *Knowledge and Human Interests* (Cambridge, UK: Polity, 1987).
80. Adorno, *Negative Dialectics*, p. 349.
81. Habermas, *Knowledge and Human Interests*.
82. Wheeler, "Antisemitism as Distorted Politics," p. 134.
83. J. Habermas, "On Systematically Distorted Communication," *Inquiry* 13: 205–8.
84. K. Marx, "German Ideology: Economic and Philosophical Manuscripts," in *Karl Marx: Selected Writings*, ed. D. McLellan (Oxford: Oxford University Press, 1977).
85. J. Janousek, "On the Marxian Concept of Praxis," in *The Context of Social Psychology: A Critical Assessment* (New York: Academic, 1972), p. 279.
86. R. Bernstein, *Beyond Objectivism and Relativism: Science, Hermeneutics and Praxis* (Oxford: Basil Blackwell, 1983), p. 187.
87. H. Gadamer, *Truth and Method* (London: Sheed and Ward, 1979), p. 275.
88. Walzer, *Interpretation and Social Criticism*, pp. 28–32.
89. Ibid.
90. J. Habermas, *Communication and the Evolution of Society* (Boston: Beacon, 1979), p. 23.
91. J. Habermas, *The Theory of Communicative Action*, vol. 2, *Life World and System: A Critique of Functionalist Reason* (Boston: Beacon, 1987), p. 383.

92. Habermas, *Communication and the Evolution of Society*, p. 23.
93. Habermas, *Theory of Communicative Action*; Honneth, "Social Dynamics of Disrespect," pp. 255–69.
94. T. Schroyer, *The Critique of Domination* (New York: Braziller, 1973), pp. 132–68.
95. L. Wittgestein, *On Certainty* (New York: Harper Torchbooks, 1969), p. 81.
96. Ibid.
97. Wittgenstein, *Philosophical Investigations*, p. 19.
98. H. Marcuse, *Eros and Civilization: A Philosophical Inquiry into Freud* (New York: Beacon, 1974); H. Marcuse, "The Obsolescence of the Freudian Concept of Man," in *Five Lectures* (Boston: Beacon, 1970), pp. 44–61; A. Honneth, and H. Joas, *Social Action and Human Nature* (New York: Cambridge University Press, 1988); A. Honneth, "Democracy as Reflexive Cooperation: John Dewey and the Theory of Democracy Today," *Political Theory* 26(6): 763–83, citing J. Dewey, *Human Nature and Conduct*, part IV (Conclusion); E. Fromm, *The Sane Society* (New York: Rinehart, 1955); E. Fromm, *Escape from Freedom* (New York: Rinehart, 1955).
99. Horkheimer, "Traditional and Critical Theory."
100. E. Fromm, "The Influence of Social Factors in Child Development," in *Yearbook of the International Erich Fromm Society*, vol. 3 (Münster: LIT-Verlag, 1992), pp. 163–65.
101. E. Fromm, "Character and the Social Process," in *Fear of Freedom* (New York: Routledge, 1942).
102. Marcuse, "Obsolescence of the Freudian Concept of Man," p. 49.
103. Marcuse, *One Dimensional Man*; Marcuse, *Eros and Civilization*.
104. M. Horkheimer, and T. Adorno, *The Dialectics of Enlightenment* (London: Verso, 1947).
105. Habermas, *Knowledge and Human Interests*.
106. Ibid.
107. Fromm, "Character and the Social Process."
108. Habermas, *Knowledge and Human Interests*, p. 159.
109. A. Feenberg, *Lukacs, Marx and the Sources of Critical Theory* (Totowa, NJ: Rowman and Littlefield, 1981), pp. 9–11.
110. J. Habermas, *The Theory of Communicative Action*, vol. 1, *Reason and Rationalization of Society* (Boston: Beacon, 1981).
111. Jay, *Dialectical Imagination*.
112. S. Bronner, *Of Critical Theory and Its Theorists* (New York: Routledge, 1993), Introduction.
113. Ibid.
114. L. Zanetti, "Advancing Praxis: Connecting Critical Theory with Practice in Public Administration," *American Review of Public Administration* 27 (2): 148–67.
115. King, "Healing the Scholarship/Practice Wounds," p. 169.

Notes to Chapter 5

1. D. Riesman, *Thorstein Veblen: A Critical Interpretation* (New York: Scribner's, 1960).
2. M. Crozier, "The Frankfurt School," in *Social Theory*, ed. P. Beilharz (North Sydney, Australia: Allen and Unwin, 1991), pp. 90–98; D. Held, *Introduction to Critical Theory: Horkheimer to Habermas* (Berkeley: University of California, 1980).

3. J. Bohman, "Critical Theory as Metaphilosophy," *Metaphilosophy* 21(3): 239–52.
4. H. Marcuse, *Eros and Civilization: A Philosophical Inquiry into Freud* (New York: Beacon, 1974).
5. M. Jay, "Some Recent Developments in Critical Theory," *Berkeley Journal of Sociology* 18: 27–44; M. Jay, *The Dialectical Imagination: A History of the Frankfurt School and the Institute of Social Research 1923–1950* (Toronto: Little, Brown, 1973).
6. J. Habermas, *The Theory of Communicative Action*, vol. 1, *Reason and Rationalization of Society* (Boston: Beacon, 1981).
7. L. Dennard, "The Democratic Potential in the Transition of Postmodernism," *American Behavioral Scientist* 41 (1): 148–62.
8. Ibid.; C. King, "Healing the Scholarship/Practice Wounds," *Administrative Theory and Praxis* (20) (2): 159–71.
9. T. Veblen, *The Place of Science in Modern Civilization and Other Essays* (New York: Russell and Russell, 1961), pp. 129–30.
10. Ibid., p. 132.
11. A. Sementelli, and C.F. Abel, "Recasting Critical Theory: Veblen, Deconstruction and the Theory Practice Gap," *Administrative Theory and Praxis* 22 (3): 458–78.
12. T.W. Adorno, "Veblen's Attack on Culture," *Studies in Philosophy and Social Science* 9 (3): 389–413; H. Marcuse, "Some Social Implications of Modern Technology," *Studies in Philosophy and Social Science* 9(3): 414–39; M. Horkheimer, "The End of Reason," *Studies in Philosophy and Social Science* 9(3): 365–88.
13. T. Veblen, *The Theory of the Leisure Class: An Economic Study of Institutions* (New York: Macmillan, 1899).
14. Ibid., p. 99.
15. Ibid., p. 34.
16. Ibid., p. 15.
17. Ibid., p. 34.
18. Ibid., p. 15.
19. Ibid., pp. 218–26; T. Veblen, "Christian Morals and the Competitive System," *International Journal of Ethics* 20 (2): 168–85.
20. H. Peukert, "On the Origins of Modern Evolutionary Economics: The Veblen Legend After 100 Years," *Journal of Economic Issues* 35 (3): 544.
21. D. Smith, *The Chicago School: A Liberal Critique* (London: Macmillan Education, 1988).
22. Peukert, "On the Origins of Modern Evolutionary Economics," p. 543.
23. Ibid., p. 543–52, note 10.
24. A. Smith, *The Wealth of Nations* (Chicago: University of Chicago Press, 1976).
25. D. North, *Institutions, Institutional Change, and Economic Performance* (Cambridge: Cambridge University Press, 1996).
26. T. Veblen, *The Theory of Absentee Ownership* (New York: Sentry Press, 1964).
27. T. Veblen, *The Theory of Business Enterprise* (New York: Augustus M. Kelley, 1965), ch. 5–6.
28. T. Veblen, *The Engineers and the Price System* (New Brunswick, NJ: Transaction, 1990).
29. H. Peukert, "On the Origins of Modern Evolutionary Economics," p. 543.
30. Veblen, "Christian Morals and the Competitive System," pp. 168–85.

31. S. Mestrovic, *Durkheim and Postmodern Culture* (New York: Aldine de Gruyter, 1992), p. 98.
32. W. Samuels, "Reflections on the Intellectual Context and Significance of Thorstein Veblen," *Journal of Economic Issues* 29 (3): 920.
33. Veblen, *Theory of the Leisure Class*, pp. 15–16, 270; T. Veblen, "The Instinct of Workmanship and the Irksomeness of Labor," *American Journal of Sociology* 4 (2): 187–201.
34. Veblen, *Theory of the Leisure Class*, pp. 218–26; Veblen, "Christian Morals and the Competitive System," pp. 168–85.
35. Samuels, "Reflections on the Intellectual Context and Significance of Thorstein Veblen," p. 918.
36. Ibid., pp. 915–23.
37. Ibid.
38. T. Veblen, *The Theory of the Business Enterprise* (New York: Random House, 1904); T. Veblen, *Absentee Ownership: Business Enterprise in Recent Times: The Case of America* (Boston, MA: Beacon, 1967).
39. Veblen, *Theory of the Leisure Class*, p. 195.
40. Ibid.
41. T. Veblen, "Why Is Economics Not an Evolutionary Science?" *Quarterly Journal of Economics* 12 (4): 373–97.
42. Ibid., p. 395.
43. Ibid., p. 377.
44. Ibid., p. 376.
45. T. Veblen, *The Instinct of Workmanship and the State of the Industrial Arts* (New York: Augustus M. Kelley, 1964), p. 18.
46. Ibid., p. 24.
47. Veblen, *Theory of the Leisure Class*, pp. 11–12, p. vii.
48. Ibid., *Theory of the Leisure Class*, p. 13;
49. T. Veblen, "Imperial Germany and the Industrial Revolution," in ed. M. Lerner, *The Portable Veblen* (New York: Viking, 1968).
50. Veblen, *Instinct of Workmanship and the State of the Industrial Arts*, p. 30.
51. M. Tool, "A Neoinstitutional Theory of Social Change in Veblen's *The Theory of the Leisure Class*," in *The Leisure Class and Sovereignty: The Centenary of the Founding of Institutional Economics*, ed. W.J. Samuels (New York: Routledge, 1986).
52. Ibid.
53. Ibid.
54. Ibid., pp. 104–5; M. Tool, *Pricing, Valuation, and Systems: Essays in Neoinstitutional Economics* (Brookfield, VT: Edward Elgarpp, 1995), 197–211.
55. T. Veblen, "The Preconceptions of the Economic Science: Part II," in *The Place of Science in Modern Civilization* (New York: B.W. Huebsch, 1919).
56. T. Veblen, "The Socialist Economics of Karl Marx," in *The Place of Science in Modern Civilization* (New York: B.W. Huebsch, 1919), p. 416.
57. T. Veblen, "Why Is Economics Not an Evolutionary Science?" in *The Place of Science in Modern Civilization* (New York: B.W. Huebsch, 1919), p. 74.
58. Veblen, "Preconceptions of the Economic Science, pp. 114–16.
59. Veblen, "Why Is Economics Not an Evolutionary Science?"
60. M. Lerner, ed. *The Portable Veblen* (New York: Viking, 1968), p. 19.
61. Veblen, *The Place of Science in Modern Civilization and Other Essays.*

NOTES TO CHAPTER 5

62. Veblen, *Theory of Business Enterprise*; Veblen, "Why Is Economics Not an Evolutionary Science?"
63. Veblen, *The Place of Science in Modern Civilization and Other Essays,* p. 129.
64. Ibid., p. 130.
65. Ibid.
66. Ibid., p. 132.
67. Ibid., p. 134.
68. J. Habermas, *Knowledge and Human Interests* (Cambridge, UK: Polity, 1987); Marcuse, *Eros and Civilization*.
69. Habermas, *Knowledge and Human Interests*.
70. Veblen, *Theory of the Leisure Class*.
71. G. Hegel, *The Philosophy of Hegel*, ed. C. Friedrich (New York: Modern Library, 1954).
72. H. Marcuse, *Reason and Revolution* (Boston: Beacon, 1941).
73. Habermas, *Knowledge and Human Interests*.
74. J. Habermas, *Communication and the Evolution of Society* (Boston: Beacon, 1979); Habermas, *Theory of Communicative Action*.
75. Veblen, *Theory of Business Enterprise* (New York: Random House, 1904).
76. Ibid., p. 308.
77. Ibid.
78. Ibid., p. 374.
79. Veblen, *Theory of Business Enterprise*, (New York: Random House, 1904), preface.
80. Riesman, *Thorstein Veblen*, pp. 152–54.
81. Habermas, *Knowledge and Human Interests*.
82. D. Elazar, *Exploring Federalism* (University: University of Alabama Press, 1987).
83. Jay, "Some Recent Developments in Critical Theory," pp. 27–44; Jay, *The Dialectical Imagination*.
84. M. Horkheimer, and T. Adorno, *Dialectic of Enlightenment* (New York: Herder and Herder, 1972).
85. H. Marcuse, *One Dimensional Man* (Boston: Beacon, 1964).
86. L. Zanetti, "Advancing Praxis: Connecting Critical Theory with Practice in Public Administration," *American Review of Public Administration* 27 (2): 148–67.
87. L. Berlin, "Two Concepts of Liberty," in *Four Essays on Liberty* (London: Oxford University Press, 1969).
88. T. McCarthy, "Jurgen Habermas," in *The Encyclopedia of Philosophy*, ed. P. Edwards (New York: Macmillan, 1967), p. 227–28.
89. Zanetti, "Advancing Praxis," pp. 148–67.
90. King, "Healing the Scholarship/Practice Wounds," pp. 159–71; Zanetti, "Advancing Praxis," pp. 148–67.
91. King, "Healing the Scholarship/Practice Wounds," p. 168.
92. B. Hall, "Introduction," in *Voices of Change*, ed. C. Park et al. (Toronto: OISE, 1993).
93. D. Baldwin, "Power Analysis and World Politics: New Trends Versus Old Tendencies," *World Politics* 31: 167–90.
94. C. Abel, and J. Oppenheimer, "Liberating the Industrious Tailor: The Case for Ideology and Instrumentalism in the Social Sciences," *Political Methodology* 8 (11): 39–59.

95. Ibid.
96. Ibid.
97. Ibid.
98. Ibid.
99. Zanetti, "Advancing Praxis," pp. 148–67.
100. King, "Healing the Scholarship/Practice Wounds," p. 169.
101. Ibid., p. 188.
102. Ibid., p. 160.
103. Ibid.
104. Dennard, "Democratic Potential in the Transition of Postmodernism," pp. 148–62.

Notes to Chapter 6

1. A. Sementelli, and C.F. Abel, "Recasting Critical Theory: Veblen, Deconstruction and the Theory Practice Gap," *Administrative Theory and Praxis* 22 (3): 458–78.
2. S. Clegg, and D. Dunkerly, *Organization, Class and Control* (London: Routledge and Kegan Paul, 1980).
3. J. Forester, *Critical Theory and Organizational Analysis* (London: Sage, 1983); J. Habermas, *The Theory of Communicative Action* (Boston: Beacon, 1984).
4. D. Mumby, "Communication and Power in Organizations," in *Discourse, Ideology and Domination*, ed., N.J. Nicholson (Norwood, NJ: Ablex, 1988).
5. J. Habermas, *Toward a Rational Society: Student Protest, Science and Politics* (Boston: Beacon, 1980).
6. D. Schon, *The Reflective Practitioner* (New York: Basic Books, 1983).
7. H. Marcuse, *Eros and Civilization* (New York: Vintage, 1985).
8. C.B. Macpherson, *The Life and Times of Liberal Democracy* (Oxford: Oxford University Press, 1977).
9. J. Newton, "Theoretizing Subjectivity on Organizations: The Failure of Foucaultian Studies," *Organization Studies* 19 (8): 414–47.
10. T. Adorno, *Prisms* (London: Neville Spearman, 1978).
11. R. Dahl, "The Concept of Power," *Behavioral Science* 2: 201–15.
12. H. Arendt, *On Violence* (London: Allen Lane, 1970); S. Lukes, *Power: A Radical View* (London: Macmillan, 1974), p. 28.
13. N. Poulantzas, *State, Power, Socialism* (London: New Left Books, 1979) p. 147.
14. P. Bachrach, and M. Baratz, "Two Faces of Power," *American Political Science Review* 56: 947–52.
15. S. Lukes, *Power: A Radical View* (London: Macmillan, 1974), p. 23.
16. H. Marcuse, *Five Lectures* (Boston: Beacon Press, 1970).
17. J. Benson, "Innovation and Crisis in Organizational Analysis," in *Organizational Analysis*, ed. J. Benson (London: Sage, 1977).
18. S. Deetz, "Disciplinary Power in the Modern Organization," in ed. *Critical Management Studies*, M. Alvesson and H. Willmott (London: Sage, 1992).
19. Habermas, *Theory of Communicative Action.*
20. B. Fay, *Critical Social Science* (Cambridge, UK: Polity, 1987); E. Fromm, *To Have or To Be?* (New York: Harper, 1976); Habermas, *Theory of Communicative Action*; M. Horkheimer and T. Adorno, *The Dialectics of Enlightenment* (London: Verso, 1947).

NOTES TO CHAPTER 6

21. G. Rose, *The Melancholy Science: An Introduction to the Thought of Theodor W. Adorno* (London: Macmillan, 1978).
22. M. Jay, *The Dialectical Imagination* (Boston: Little, Brown, 1973); P. Slater, *Origin and Significance of the Frankfurt School* (New York: Routledge, 1977).
23. Marcuse, *Five Lectures*, p. 88.
24. Ibid., pp. 1–2.
25. A. Wellmer, "Reason, Utopia and the Dialectics of Enlightenment," in *Habermas and Modernity*, ed. J. Bernstein (Cambridge: Polity, 1985).
26. Marcuse, *Five Lectures*, pp. 1–2.
27. Habermas, *Theory of Communicative Action*.
28. T. Schroyer, *The Critique of Domination: The Origins and Development of Critical Theory* (Boston: Beacon, 1975).
29. Fromm, *To Have or To Be?*
30. Fay, *Critical Social Science*.
31. Horkheimer and Adorno, *Dialectics of Enlightenment*; H. Marcuse, *One Dimensional Man* (Boston: Beacon, 1964).
32. L. Wittgenstein, *Philosophical Investigations* (New York: Macmillan, 1953).
33. Ibid.
34. J. March, and H. Simon, *Organizations* (New York: Wiley, 1958).
35. H. Kaufman, "Administrative Decentralization and Political Power," *Public Administration Review* 29: 3–15; M. Landau, "Redundancy, Rationality and the Problem of Duplication and Overlap," *Public Administration Review* 19: 79–88; A. Wildavsky, *Speaking Truth to Power* (Boston: Little, Brown, 1979).
36. B. Adam, *The Survival of Domination: Inferiorization and Everyday Life* (New York: Elsevier North-Holland, 1978); C. Gould, "Private Rights and Public Virtues: Women, the Family, and Democracy," in *Beyond Domination: New Perspectives on Women and Philosophy*, ed. C.G. Gould (Totowa, NJ: Rowman and Allanheld, 1984); J. Hearn, *The Gender of Oppression: Men, Masculinity, and the Critique of Marxism* (Brighton: Wheatsheaf Books, 1987); A. Memmi, *Dominated Man: Notes Towards a Portrait* (Boston: Beacon, 1968).
37. C. Abel, and F. Marsh, *Political Trials: Criticisms and Justifications* (Westport, CT: Greenwood Press, 1993).
38. Ibid.
39. A. Sementelli, and C.F. Abel, "Recasting Critical Theory: Veblen, Deconstruction and the Theory Practice Gap," *Administrative Theory and Praxis* 22 (3): 458–78.
40. R. May, *Power and Innocence* (New York: Norton, 1972), pp. 121–22.
41. T. Veblen, *The Theory of the Leisure Class* (New York: Random House, 1934), p. 15.
42. T. Veblen, "The Instinct of Workmanship and the Irksomeness of Labor," *American Journal of Sociology* 4 (2): 187–201.
43. T. Veblen, *Theory of the Leisure Class*, pp. 1–21.
44. May, *Power and Innocence*.
45. Ibid., pp. 143–44.
46. Veblen, *Theory of the Leisure Class*, p. 16.
47. F. Nietzsche, *The Will to Power* (New York: Vintage, 1968), p. 340.
48. P. Tillich, *Love, Power and Justice* (New York: Oxford University Press, 1954), p. 39.

49. H. Ansbacher, and R. Ansbacher, eds. *The Individual Psychology of Alfred Adler* (New York: Harper and Row, 1956).
50. M. Foucault, *The History of Sexuality*, vol. 1, *An Introduction* (London: Allen Lane, 1979), p. 92.
51. Ibid.
52. Ibid.
53. T. Veblen, *The Theory of the Business Enterprise* (New York: Random House, 1904), p. 400; T. Veblen, *Absentee Ownership: The Case of America* [1923] (Boston: Beacon Press, 1967).
54. M. Foucault, *Discipline and Punish: The Birth of the Prison* (New York: Pantheon, 1977).
55. Foucault, *History of Sexuality*, pp. 92–93.
56. Ibid., p. 93.
57. M. Foucault, *Power/Knowledge* (New York: Pantheon, 1980), p. 131.
58. Ibid.
59. T. Veblen, *The Place of Science in Modern Civilization* (New York: Random House, 1932).
60. T. Veblen, *The Instinct of Workmanship* (New York: Augustus Kelley, 1964).
61. Foucault, *History of Sexuality*.
62. Foucault, *Power/Knowledge*, p. 93.
63. Horkheimer and Adorno, *Dialectics of Enlightenment*, p. 222; Marcuse, *One Dimensional Man*.
64. H. Marcuse, "Repressive Tolerance," in *A Critique of Pure Tolerance*, ed. R. Wolf et al. Boston, MA: Beacon, 1965), p. 116.
65. M. Jay, *The Dialectical Imagination* (Boston: Little, Brown, 1973), p. 279.
66. Veblen, *Place of Science in Modern Civilization*, pp. 74–75.
67. Veblen, *Theory of the Leisure Class*, pp. 198–211.

Notes to Chapter 7

1. F. Marini, ed. *Toward a New Public Administration* (Scranton, PA: Chandler, 1971); J. Rohr, *To Run a Constitution* (Lawrence: University Press of Kansas, 1986); M. Spicer, and L. Terry, "Legitimacy, History and Logic: Public Administration and the Constitution," *Public Administration Review* 53: 239–45; L. Terry, *Leadership of Public Bureaucracies: The Administrator as Conservator* (Thousand Oaks, CA: Sage, 1995); V. Thompson, *Without Sympathy or Enthusiasm: The Problem of Administrative Compassion* (Tuscaloosa, AL: University of Alabama Press, 1975).
2. G. Roth, and C. Wittich, eds. *Max Weber Economy and Society* (Berkeley: University of California Press, 1978), p. 223.
3. T. Veblen, *The Vested Interests and the Common Man* (New York: Sentry Press, 1964), p. 57.
4. J. Habermas, *The Theory of Communicative Action* (Boston: Beacon, 1984).
5. H. Marcuse, *Five Lectures* (Boston: Beacon, 1970), pp. 1–2.
6. A. Wellmer, "Reason, Utopia and the Dialectics of Enlightenment," in *Habermas and Modernity*, ed. J. Bernstein (Cambridge, UK: Polity, 1985).
7. Habermas, *Theory of Communicative Action*.
8. A. Honneth, *The Struggle for Recognition* (Cambridge, MA: Polity, 1994).
9. S. Wolin, "Political Theory as a Vocation," *American Political Science Review* 63: 1062–82.

10. A. Benzaquen, "Thought and Utopia in the Writings of Adorno, Horkheimer, and Benjamin," *Utopian Studies* (Spring): 150.

11. T. Adorno, "Sociology and Empirical Research," in *Critical Sociology*, ed. P. Connerton (New York: Penguin, 1976), p. 248; Habermas, *Theory of Communicative Action*; M. Horkheimer, *Critical Theory* (New York: Seabury, 1972), p. 213, 233; C. King, "Healing the Scholarship/Practice Wounds," *Administrative Theory and Praxis* 20 (2): 159–71; L. Zanetti, "Advancing Praxis: Connecting Critical Theory with Practice in Public Administration," *American Review of Public Administration* 27 (2): 148–67.

12. Plato, *The Republic* (New York: Oxford University Press, 1945), p. 282.

13. M. Walzer, "Flight from Philosophy," *New York Review of Books* 36 (1): 42–44.

14. J. Rawls, *A Theory of Justice* (Cambridge, MA: Harvard University Press, 1971), p. 253.

15. B. Barber, *Strong Democracy: Participation in Politics for a New Age* (Berkeley: University of California Press, 1984).

16. W. Kymlicka, *Contemporary Political Philosophy* (New York: Oxford University Press, 1990), p. 208.

17. Ibid.

18. M. Sandel, *Liberalism and It's Critics* (Oxford: Basil Blackwell, 1984), p. 206.

19. M. Sandel, *Liberalism and the Limits of Justice* (Cambridge, UK: Cambridge University Press, 1982), p. 258.

20. Ibid., p. 21.

21. Horkheimer, *Critical Theory*.

22. Habermas, *Theory of Communicative Action*.

23. A. Honneth, and J. Farrell, "Recognition and Moral Obligation," *Social Research* 64 (1): 22.

24. Honneth, *Struggle for Recognition*.

25. J. Dunn, ed. *The Economic Limits to Modern Politics* (Cambridge, UK: Cambridge University Press, 1990), p. 6; A. Hamlin, and P. Pettit, eds., *The Good Polity: Normative Analysis of the State* (Oxford: Basil Blackwell, 1989); Rawls, *A Theory of Justice*.

26. B. Fay, *Social Theory and Political Practice* (London: Allen and Unwin, 1975), pp. 92–110.

27. S. White, "Reason and Authority in Habermas: A Critique of the Critics," *American Political Science Review* 74: 1007–17.

28. J. Ibanez-Noe, "The Dialectic of Emancipation and Power and the Nihilistic Character of Modernity," *CLIO* 28 (2): 150.

29. B. Shaw, "Reason, Nostalgia, and Eschatology in the Critical Theory of Max Horkheimer," *Journal of Politics* 47 (1): 166.

30. M. Horkheimer, *Dawn and Decline: Notes 1926–1931 and 1950–1969* (New York: Seabury, 1978); Shaw, "Reason, Nostalgia, and Eschatology in the Critical Theory of Max Horkheimer," p. 167.

31. M. Horkheimer, *Eclipse of Reason* (New York: Seabury, 1974), p. 149.

32. T. Adorno, "Repressive Tolerance," in *Critique of Pure Tolerance*, ed. H. Marcuse (Boston: Beacon, 1965), p. 105.

33. T. Adorno, *Negative Dialectics* (London and New York: Routledge, 1973), p. 150.

34. A. Macintyre, *Herbert Marcuse* (New York: Viking, 1970), pp. 99–105.

35. H. Bredekamp, "From Walter Benjamin to Carl Schmitt, via Thomas Hobbes," *Critical Inquiry* 25 (2), p. 247.

36. G. Hartman, "Benjamin in Hope," *Critical Inquiry* 25 (2): 344–49.
37. J. Habermas, *Between Facts and Norms: Contributions to a Discourse Theory of Law and Democracy* (Cambridge, UK: Polity, 1996).
38. Honneth and Farrell, "Recognition and Moral Obligation," p. 19.
39. B. Fay, *Critical Social Science* (Cambridge, UK: Polity, 1987); E. Fromm, *To Have or To Be?* (New York: Harper, 1976); Habermas, *Theory of Communicative Action*; Honneth and Farrell, "Recognition and Moral Obligation," pp. 16–51; M. Horkheimer, and T. Adorno, *Dialectic of Enlightenment* (London: Verso, 1947).
40. Shaw, "Reason, Nostalgia, and Eschatology in the Critical Theory of Max Horkheimer," p. 180.
41. M. Alvesson, and H. Willmott, "On the Idea of Emancipation in Management and Organization Studies," *Academy of Management Review* 17 (3): 444.
42. Habermas, *Theory of Communicative Action*; J. Habermas, "Crossing Globalizations Valley of Tears," *New Left Review* (Fall): 51–57.
43. Honneth, *Struggle for Recognition*.
44. J. Forester, *Critical Theory and Organizational Analysis* (London: Sage, 1983); Habermas, *Theory of Communicative Action*.
45. Wellmer, "Reason, Utopia and the Dialectics of Enlightenment."
46. Habermas, *Between Facts and Norms*, pp. 179, 306.
47. Horkheimer, *Critical Theory*.
48. Fay, *Social Theory and Political Practice*.
49. L. Armour, "Economics and Social Reality: Professor O'Neal and the Problem of Culture," *International Journal of Social Economics* 22 (9): 79–89.
50. W. French, and C. Bell, *Organization Development: Behavior Science Interventions for Organizational Improvement* (Englewood Cliffs, NJ: Prentice-Hall, 1995); T. Veblen, *The Theory of the Leisure Class: An Economic Study of Institutions* (New York: Macmillan, 1934).
51. S. Zuboff, *In the Age of the Smart Machine* (New York: Basic Books, 1988).
52. *Reynolds v. United States* 98 U.S. 145 (1878).
53. B. Adam, *The Survival of Domination: Inferiorization and Everyday Life* (New York: Elsevier North-Holland, 1978); C. Gould, "Private Rights and Public Virtues: Women, the Family, and Democracy," in *Beyond Domination: New Perspectives on Women and Philosophy*, ed. C.G. Gould (Totowa, NJ: Rowman and Allanheld, 1984); J. Hearn, *The Gender of Oppression: Men, Masculinity, and the Critique of Marxism* (Brighton: Wheatsheaf Books, 1987); A. Memmi, *Dominated Man: Notes Towards a Portrait* (Boston: Beacon, 1968).
54. *Yoder v. Wisconsin* 406 U.S. 208 (1972).
55. B. Barnes, "Status Groups and Collective Action," *Sociology* 26 (2): 259–70.
56. Ibid., p. 262.
57. Veblen, *Theory of the Leisure Class*.
58. Horkheimer, *Eclipse of Reason*, pp. 101–2.
59. Honneth and Farrell, "Recognition and Moral Obligation," p. 22.
60. J. Habermas, "On the Pragmatic, the Ethical, and the Moral Employments of Practical Reason," in *Habermas, Justification and Application*. (Boston: MIT, 1995), p. 1.
61. Ibid., pp. 1–18.
62. Honneth and Farrell, "Recognition and Moral Obligation," p. 22.
63. Honneth, *Struggle for Recognition*.
64. A. Sementelli, and C.F. Abel, "Recasting Critical Theory: Veblen, Deconstruction and the Theory Practice Gap," *Administrative Theory and Praxis* 22 (3): 458–78.

65. Veblen, *Vested Interests and the Common Man*, p. 57.
66. Ibid.
67. Ibid., p. 117.
68. Ibid., pp. 119–20.
69. Ibid., p. 119.
70. Alvesson and Willmott, "On the Idea of Emancipation in Management and Organization Studies," p. 432–64.
71. T. Veblen, *The Theory of the Business Enterprise* (New York: Random House, 1904); T. Veblen, *The Place of Science in Modern Civilization and Other Essays* (New York: Russell and Russell, 1961).
72. Veblen, *Place of Science in Modern Civilization*, p. 395.
73. Ibid., p. 377.
74. Ibid., p. 376.
75. T. Veblen, *The Place of Science in Modern Civilization* (New York: Random House, 1932).
76. D. Gioia, M. Schultz, and K. Corley, "Organizational Identity, Image, and Adaptive Instability," *Academy of Management Review* 25 (1): 64.
77. Veblen, *Theory of the Leisure Class*.
78. Veblen, *Vested Interests and the Common Man*, pp. 32–33.
79. Ibid., pp. 1–21.
80. F. Jameson, "The Theoretical Hesitation: Benjamin's Sociological Predecessor," *Critical Inquiry* 25 (2): 269.
81. M. Foucault, *Power/Knowledge* (New York: Pantheon, 1980), p. 131.
82. Alvesson and Willmott, "On the Idea of Emancipation in Management and Organization Studies," pp. 432–64; Foucault, *Power/Knowledge*; Gioia, Schultz, and Corley, "Organizational Identity, Image, and Adaptive Instability," pp. 63–81; Roth and Wittich, *Max Weber Economy and Society*, p. 1194; Veblen, *Place of Science in Modern Civilization*.
83. M. Bakhtin, *Speech Genres and Other Late Essays* (Austin: University of Texas Press, 1986).
84. Foucault, *Power/Knowledge*, p. 131.
85. Ibid.
86. Ibid.
87. Alvesson and Willmott, "On the Idea of Emancipation in Management and Organization Studies," pp. 432–64.
88. Veblen, *Vested Interests and the Common Man*, p. 57.

Notes to Chapter 8

1. J. Rohr, *To Run a Constitution* (Lawrence: University Press of Kansas, 1986); L. Terry, *Leadership of Public Bureaucracies: The Administrator as Conservator* (Thousand Oaks, CA: Sage, 1995).
2. A. Donagan, "Historical Explanation: The Popper-Hempel Theory Reconsidered," *History and Theory* 6: 3–26; J. Nelson, "Accidents, Laws, and Philosophic Flaws: Behavioral Explanations in Dahl and Dahrendorf," *Comparative Politics* 7: 435–57.
3. S. Verba, "Comparative Politics: Where Have We Been, Where Are We Going?" in *New Directions in Comparative Politics*, ed. H.J. Wiarda (Boulder, CO: Westview, 1985), p. 34.

4. R. Merton, *Social Theory and Social Structure* (New York: Free Press, 1957), pp. 5–6.

5. M. Foucault, *Power/Knowledge* (New York: Pantheon, 1980), p. 131.

6. M. Alvesson, and H. Willmott, "On the Idea of Emancipation in Management and Organization Studies," *Academy of Management Review* 17 (3): 432–64; Foucault, *Power/Knowledge*; D. Gioia, M. Schultz, and K. Corley, "Organizational Identity, Image, and Adaptive Instability," *Academy of Management Review* 25 (1): 63–81; G. Roth, and C. Wittich, eds. *Max Weber Economy and Society* (Berkeley: University of California Press, 1978), p. 1194; T. Veblen, *The Place of Science in Modern Civilization and Other Essays* (New York: Russell and Russell, 1961).

7. M. Bakhtin, *Speech Genres and Other Late Essays* (Austin: University of Texas Press, 1986).

8. Alvesson and Willmott, "On the Idea of Emancipation in Management and Organization Studies," pp. 432–64.

9. T. Veblen, *The Vested Interests and the Common Man* (New York: Sentry, 1964), p. 57.

10. L. Gulick, and L. Urwick, *Papers on the Science of Administration* (New York: Institute of Public Administration, 1937); F. Goodnow, *Politics and Administration* (New York: Macmillan, 1900); H. Metcalf, and L. Urwick, *Dynamic Administration: The Collected Papers of Mary Parker Follet* (London: Sir Isaac Pittman and Sons, 1941); W. Wilson, "The Study of Public Administration," *Political Science Quarterly* 2 (June): 197–222;

11. F. Marini, ed. *Toward a New Public Administration* (Scranton, PA: Chandler, 1971); Rohr, *To Run a Constitution*; H.J. Storing, "American Statesmanship: Old and New," in ed. R.A. Goldwin, *Bureaucrats, Policy Analysts, and Statesmen: Who Leads?* (Washington, DC: American Enterprise Institute for Public Policy Research, 1980), pp. 88–113; D. Waldo, "The Study of Public Administration" (New York: Random House, 1955); D. Waldo, *The Administrative State: A Study of the Political Theory of American Public Administration* (New York: Holmes and Meir, 1984); G. Wamsley, R. Bacher, C. Goodsell, P. Kronenberg, J. Rohr, C. Stivers, O. White, and J. Wolf, *Refounding Public Administration* (Newbury Park, CA: Sage, 1990).

12. S. Abbasi, "A Comparison of Private and Public Managers' Value Systems (Ph.D. diss., Mississippi State University, 1982); B. Bozeman, *All Organizations Are Public: Bridging Public and Private Organizational Theories* (San Francisco: Jossey Bass, 1989); B. Bozeman, ed., *Public Management: The State of the Art* (San Francisco: Jossey Bass, 1993); M. Murray, "Comparing Public and Private Management: An Exploratory Essay," *Public Administration Review* 34 (4): 364–71; H. Rainey, R. Backoff, and C. Levine, "Comparing Public and Private Organizations," *Public Administration Review* (March/April): 233–44; H. Rainey, *Understanding and Managing Public Organizations* (San Francisco: Jossey Bass, 1997); F. Russo, Jr., "The Case for Private Management," *Public Management* (January/February): 11–14; E. Savas, *Privatizing the Public Sector* (Chatham, NJ: Chatham House, 1982).

13. L. Keller, and M. Spicer, "Political Science and American Public Administration: A Necessary Cleft?" *Public Administration Review* 57 (3): 270; M. Whicker, R. Strickland, and D. Olshfski, "The Troublesome Cleft: Public Administration and Political Science," *Public Administration Review* 53 (6): 531–41; L. Mainzer, "Public Administration in Search of a Theory: The Interdisciplinary Delusion," *Administration and Society* 26 (3): 359–94.

14. Whicker, Strickland, and Olshfski, "Troublesome Cleft," pp. 531–41; Mainzer, "Public Administration in Search of a Theory," pp. 359–94.

15. G. Boyne, "Public and Private Management: What's the Difference?" *Journal of Management Studies* 39 (1): 97–122.

16. C. Pollit, "Clarifying Convergence Striking Similarities and Durable Differences in Public Management Reform," *Public Management Review* 4 (1): 471–92; N. Riccuci, "The 'Old' Public Management Versus the 'New' Public Management: Where Does Public Administration Fit In?" *Public Administration Review* 61 (2): 172–75; R. Wettenhall, and I. Thynne, "Public Enterprise and Privatization in a New Century: Evolving Patterns of Governance and Public Management," *Public Finance and Management* 2 (1): 1–29.

17. R. Box, G. Marshal, B. Reed, and C. Reed, "New Public Management and Substantive Democracy," *Public Administration Review* 61 (5): 608–19.

18. D. Smithburg, "Political Theory and Public Administration," *Journal of Politics* 13 (1): 59–69; Waldo, *Study of Public Administration*; Waldo, *Administrative State*.

19. Keller, and Spicer, "Political Science and American Public Administration," p. 270; J. Raadschelders, "A Coherent Framework for the Study of Public Administration," *Journal of Public Administration Research and Theory* 9 (2): 281–303.

20. J. Gaus, *Reflections on Public Administration* (Tuscaloosa: University of Alabama Press); Waldo, *Administrative State*.

21. Rohr, *To Run a Constitution*; Storing, "American Statesmanship," pp. 88–113.

22. Rohr, *To Run a Constitution*; H. Taylor, *The Statesman* (Westport, MA: Praeger, 1992).

23. E. Lee, "Political Science, Public Administration, and the Rise of the American Administrative State," *Public Administration Review* 55 (6): 538–46.

24. Whicker, Strickland, and Olshfski, "Troublesome Cleft," pp. 531–41.

25. Keller and Spicer, "Political Science and American Public Administration," p. 270.

26. Whicker, Strickland, and Olshfski, "Troublesome Cleft," p. 531.

27. Lee, "Political Science, Public Administration, and the Rise of the American Administrative State," p. 544.

28. Ibid., p. 544.

29. Keller and Spicer, "Political Science and American Public Administration," p. 271.

30. T. Lowi, *The End of the Republican Era* (Norman: University of Oklahoma Press, 1995); T. Lowi, *The End of Liberalism* (New York: Norton, 1979).

31. Terry, *Leadership of Public Bureaucracies*.

32. T. Adorno, *Prisms* (London: Neville Spearman, 1978).

33. Ibid., p. 296.

34. Ibid., pp. 296–97.

35. Lowi, *End of the Republican Era*, p. 63.

36. Ibid., p. 157.

37. Ibid., p. 215.

38. G. Rose, *The Melancholy Science: An Introduction to the Thought of Theodor W. Adorno* (London: Macmillan, 1978).

39. M. Jay, *The Dialectical Imagination* (Boston: Little, Brown, 1973); P. Slater, *Origin and Significance of the Frankfurt School* (New York: Routledge, 1977).

40. C. Goodsell, "Public Administration as Republican Ally," *Public Administration Review* 55 (5): 480.

41. Ibid.
42. D. Schoenbrod, *Power Without Responsibility: How Congress Abuses the People Through Delegation* (New Haven, CT: Yale University Press, 1993).
43. Ibid., p. 13.
44. Ibid., p. 17.
45. Ibid., p. 20.
46. L. Fisher, "Power Without Responsibility," *Public Administration Review* 55 (4): p. 384.
47. Schoenbrod, *Power Without Responsibility*, p. 67.
48. Ibid., p. 118.
49. Ibid., p. 133, citing J.L. Mashaw, "The Economics of Politics and the Understanding of Public Law 65," *Chicago-Kent Law Review* 123: 127.
50. C. Kerwin, *Rulemaking: How Government Agencies Write and Make Law* (Washington, DC: Congressional Quarterly Press, 1994); M. Spicer, "On Friedrich Hayek and Public Administration: An Argument for Discretion Within Rules," *Administration and Society* 25 (1): 46–60.
51. C. Jones, *The Presidency in a Separated System* (Washington, DC: Brookings Institution, 1994), p. 24.
52. Ibid., p. 25.
53. Ibid., p. 25.
54. S. Stehr, "Top Bureaucrats and the Distribution of Influence in Reagan's Executive Branch," *Public Administration Review* 57 (1): 75–82.
55. Jones, *Presidency in a Separated System*, p. 53.
56. Ibid.
57. Ibid., p. 101.
58. Ibid., p. 285.
59. Rohr, *To Run a Constitution*.
60. Terry, *Leadership of Public Bureaucracies*.
61. L. Terry, "Public Administration and the Theater Metaphor: The Public Administrator as Villain, Hero, and Innocent Victim," *Public Administration Review* 57 (1): 53–62.
62. M. Spicer, and L. Terry, "Legitimacy, History and Logic: Public Administration and the Constitution," *Public Administration Review* 53: 239–45; M. Spicer, and L. Terry, "Administrative Interpretation of Statutes: A Constitutional View on the 'New World Order' of Public Administration," *Public Administration Review* 56 (1): 38–47.
63. C. Stivers, "Rationality and Romanticism in Constitutional Argument," *Public Administration Review* 53 (3): 256.
64. J. Rohr, "Toward a More Perfect Union," *Public Administration Review* 53 (3): 248.
65. G. Burrell, and G. Morgan, *Sociological Paradigms and Organizational Analysis* (Brookfield, VT: Ashgate Publishing, 1993), pp. 25–28.
66. Ibid., pp. 28–32.
67. Ibid., pp. 33–35.
68. Ibid., pp. 32–33.
69. J. White, G. Adams, J. Forrester, "Knowledge and Theory Development in Public Administration: The Role of Doctoral Education and Research," *Public Administration Review* 56 (5): 441–52.
70. Ibid., p. 451.
71. C. Benjamin, "Types of Empiricism," *Philosophical Review* 51 (5): 497–502.

72. Ibid., p. 497–98.
73. Ibid., p. 499.
74. W. Heisenberg, "Paper on Uncertainty Principle," (English reprint); D. Cassidy, "Heisenberg, Uncertainty and the Quantum Revolution," *Scientific American* 266 (May): 106–12.
75. C. Darwin, "The Uncertainty Principle" *Science*, n.s., 73 [1903], (1931): 660.
76. R. Eaton, "What is the Problem of Knowledge?" *Journal of Philosophy* 20 (7): 180.
77. Ibid., p. 180.
78. Ibid., p. 181.
79. Ibid.
80. Burrell and Morgan, *Sociological Paradigms and Organizational Analysis*.
81. R. Nelson, "Behaviorism Is False," *Journal of Philosophy* 66 (14): 417.
82. Heisenberg, "Paper on Uncertainty Principle"; Cassidy, "Heisenberg, Uncertainty and the Quantum Revolution," pp. 106–12.
83. C. Darwin, "Uncertainty Principle," pp. 653–60.
84. K. Popper, "Humanism and Reason," *Philosophical Quarterly* 2 (7): 171.
85. Ibid.
86. G. Trey, *Solidarity and Difference: The Politics of Enlightenment in the Aftermath of Modernity* (Albany: State University of New York Press, 1998), p. 162.

Index

A

Action research, 39
Agency
  critical theory, 90–91, 145–48
  Evolutionary Critical Theory, 154, 156
  ontological status, 51–55
  public administration theory, 68–69, 72–73
Authoritarianism, 58–59

B

Balance wheel, 169
Behaviorism
  institutional/behavioral/hermeneutic approach, 5, 34, 38–39, 64–65
  public administration epistemology, 171–72

C

Capitalism, 95, 99–100, 108, 115, 165
Cause-and-effect approach, 103–4
Christianity, 100, 108
Civil liberty, 61
Clustering approach
  disciplinary status, 15–18
  public administration theory, 66–67
Coercive power, 149–50
Cognitive authority, 58–60
Cognitive lenses, 10–11
Collaboration, 111–12

Collectivism
  critical theory, 126, 128
  disciplinary status, 29
  Evolutionary Critical Theory, 114
  ontological status, 48–51
  public administration theory, 66–67
Comparative methodology, 26, 27–28
Conservatism
  critical theory, 128
  Evolutionary Critical Theory, 135–36
  power relationships, 128, 135–36
  theoretical, 11
Conservatorship, 169
Critical theory
  action research, 66
  case studies, 66
  concept analysis, 66
  deconstruction technique, 66, 81, 86–87, 91–92, 96, 97, 99, 142, 145
  defined, 66
  disciplinary status, 12–13
  empowerment, 66
  evolutionary paradigm
    democratic ideology, 78, 91, 92–93
    emancipation, 94
    empowerment, 93
    endogenous, 65
    intersubjective experience of good governance, 92–93
    power relationships, 94
    theory-practice gap, 93

Critical theory *(continued)*
  family resemblance approach, 79, 86, 96, 128
  hermeneutics synergy
    deconstruction technique, 86–87
    discourse analysis, 86–88
    discourse ethics, 86–88
    family resemblance approach, 86
    reflective synergy, 77–78
    social constructionism, 86–90
    social reality, 88–90
    systems of thought, 85–87
    transcendental hermeneutics, 87–88, 90, 91–92
  institutionalism synergy
    dominant relations, 84–85
    empirical theory, 83–84
    immanent critique, 83–85, 90, 91–92
    normative theory, 83
    positivism, 84
    power relationships, 82–85, 94
    praxis, 83, 85, 90, 91–92
    reflective synergy, 77–78
    social constructionism, 83–84
  intersubjective experience of good governance
    critical theory role, 77–78, 79
    evolutionary paradigm, 92–93
    public administration application, 90, 92
  intersubjective pluralism, 78
  objectives, 66
  ontological status, 48
  overview, 6–9
  public administration application, 90–94
    agency, 90–91
    deconstruction technique, 91–92
    institutionalism, 90–91
    intersubjective experience of good governance, 90, 92
    public administration subject matter, 90–91

Critical theory
  public administration application *(continued)*
    scientific method, 90
    social constructionism, 91
    social reality, 91
    transition points, 90
  public administration legitimacy, 166
  reflective synergy, 77–78, 80, 81
  role of, 77–90
  social reality, 79, 80, 88–90, 91
  social science synergy
    deconstruction technique, 81
    empirical theory, 80, 81
    negative dialectics, 81–82, 90, 91–92
    positivism, 80–81
    reflective synergy, 77–78, 80, 81
    scientific method, 81–82
    social constructionism, 81–82
    social reality, 80
  social science theory
    metatheory, 78
    research programs, 78
  synergistic agent, 6, 79-80
  synergistic development, 64-65, 77–78
  See also Evolutionary Critical Theory; Good society; Power relationships; Teleology; Veblen, Thorstein

D

Deconstruction technique
  critical theory, 66, 81, 86–87, 91–92, 96, 97, 99, 114, 142, 145
  disciplinary status, 16–18
  See also specific technique
Democratic ideology
  critical theory, 7–8, 96, 110–11, 112, 113–14
  evolutionary paradigm, 78, 91, 92–93
  good society, 141, 144

Democratic ideology
  critical theory *(continued)*
    power relationships, 121, 127
  disciplinary status, 15
  Evolutionary Critical Theory, 118, 119
  ontological status, 50
  public administration legitimacy, 165
Determinism
  critical theory, 96–97, 102–3
  Evolutionary Critical Theory, 117–19
  ontological status, 52–55
  public administration theory, 72–73
  sociological determinism, 52–53, 54–55
Dialectical method, 105–6, 157
Disciplinary matrix
  evolution of, 12–13
  intersubjective experience of good governance, 5, 28, 31–32
  orthodoxy utilization, 14, 23, 28
  public administration identity, 162–64
  theoretical estrangement, 12–13
Disciplinary paradigms
  orthodoxy utilization, 14
  subparadigms
    cognitive approach, 30–31
    ethical approach, 31–32
Disciplinary status
  clustering approach
    deconstruction technique, 16–18
    interconnections, 18
    intersubjective experience of good governance, 16
    postmodernism, 16–18
    public administration defined, 16–18, 25
    research review, 15–16
  conclusions, 32–33
  family resemblance approach
    conceptualization disputes, 24–25
    defined, 22–23
    discourse theory, 20–22
    intersubjectivity, 19–22
    linguistic interconnections, 18–19

Disciplinary status
  family resemblance approach *(continued)*
    orthodoxy utilization, 21–23
    power relationships, 20
    public administration defined, 19–25, 29–30
    public administration identity, 163
    public administration origin, 23–24
    redefinition approach, 20–22, 29–30
    research review, 20–21
    socially negotiated complexes, 19–20, 22–23
  intersubjective experience of good governance
    clustering approach, 16
    cognitive approach subparadigm, 30–31
    collectivism, 29
    comparative methodology, 26, 27–28
    disciplinary matrix, 28, 31–32
    ethical approach subparadigm, 31–32
    orthodoxy utilization, 14, 25
    paradigms, 30–32
    political science framework, 27–28, 32
    politics-administration distinction, 25–26, 27–28
    popular sovereignty, 25–26
    positivism, 28
    power relationships, 28
    public administration defined, 25–32
    public administration origin, 25–32
    rationality, 27
    redefinition approach, 28–29
    research publications, 29–30
    scientific method, 26, 27, 31
    social science research, 26–28
    subparadigms, 30–32
    theoretical estrangement, 12

Disciplinary status *(continued)*
  intersubjectivity
    family resemblance approach,
      19–22
    theoretical estrangement, 10, 11
  orthodoxy utilization
    democratic ideology, 15
    disciplinary matrix, 14, 23, 28
    family resemblance approach,
      21–23
    heuristic devices, 14
    ideal form, 13–14
    intersubjective experience of good
      governance, 14, 25
    paradigms, 14
    politics-administration distinction,
      14
    scientific method, 14
  overview, 3, 13, 160–62
  scientific method, 14, 26, 27, 31, 34
  theoretical estrangement
    abstractions, 10, 11
    cognitive lenses, 10–11
    critical theory, 12–13
    disciplinary matrix evolution,
      12–13
    Evolutionary Critical Theory, 13
    heuristic devices, 10–11
    identity crisis, 9, 12–13, 30
    intersubjective experience of good
      governance, 12
    intersubjectivity, 10, 11
    ontological status, 12–13
    postmodernism, 11
    rationality, 11
    situational theory, 10
    theoretical conservatism, 11
    theory-practice gap, 11, 13
  See also Ontological status
Discourse analysis
  critical theory, 86–88
  Evolutionary Critical Theory, 98, 101,
    105–6
Discourse ethics, 86–88, 110–11, 149

Discourse theory, 20–22
Dominant relations
  critical theory, 84–85, 98–103
    good society, 141–44, 145–48, 149
    power relationships, 122–23,
      124–29
  Evolutionary Critical Theory, 105,
    106–7, 110–19
    good society, 151–54
    power relationships, 131–36
  See also Power relationships
Drift
  business enterprise, 107, 152
  customs/practices, 19–22, 77, 152,
    155, 161
  emancipation, 132–36
  Evolutionary Critical Theory, 132–36
  language game, 19–22, 77
  power relationships, 132–36
  praxis, 101
  public administration legitimacy, 166,
    169

E

Economic ideology
  capitalism, 95, 99–100, 108, 115, 165
  critical theory, 95, 98–100, 102, 165
  Evolutionary Critical Theory, 103–7,
    108, 115
Emancipation
  critical theory
    conceptualization, 121, 122,
      125–26
    conceptualization paradoxes,
      126–30, 133–36
    evolutionary paradigm, 94
    good society, 141–50
    reconstruction technique, 107
    social constructionism, 96–97,
      108–10
  Evolutionary Critical Theory
    defined, 121–22, 130–36
    drift, 132–36

Emancipation
  Evolutionary Critical Theory
    (continued)
    good society, 150–57
    theory-practice gap, 114, 119
  ontological status, 48–49, 58, 61–62
  See also Power relationships
Emotional security, 144, 149
Empirical theory
  critical theory, 80, 81, 83–84
  Evolutionary Critical Theory, 113
  ontological status, 34, 40, 55–56
  public administration
    epistemology, 170–72
    theory, 64–65, 74–75
Empowerment
  critical theory, 66, 96, 113–14, 128, 143–44
  Evolutionary Critical Theory, 133
  hermeneutics, 66
  institutionalism, 66
  social science research, 66
Endogenous evolution
  critical theory reconstruction, 65, 95, 98, 103–7
  power relationships, 98–100, 102–3, 104–5, 121, 133–34
  public administration legitimacy, 169
Endogenous forces, 148, 154, 161
Epistemology, 170–72
Equality, 61
Evolutionary Critical Theory
  cause-and-effect approach, 103–4
  conclusions, 119–20
  critical theory nature
    capitalism, 95, 99–100, 108, 115
    Christianity, 100, 108
    collaboration, 111–12
    collectivism, 114
    deconstruction technique, 96, 97, 99, 114
    democratic ideology, 96, 110–11, 112, 113–14, 118, 119
    determinism, 96–97, 102–3

Evolutionary Critical Theory
  critical theory nature *(continued)*
    dialectical method, 105–6
    discourse analysis, 98, 101, 105–6
    discourse ethics, 110–11
    dominant relations, 98–103
    economic ideology, 95, 98–100, 102
    empirical theory, 113
    empowerment, 96, 113–14
    habits of thought, 98, 100, 101–2
    ideal speech situation, 108–9, 110–11, 112, 113–14
    immanent critique, 95, 96, 97, 99–100, 102–4, 108–9
    instrumentalism, 97, 112–14, 116
    latent functions, 105
    materialism, 108
    metaphysical approach, 103–4
    natural laws, 97, 104
    natural rights, 100
    negative dialectics, 98, 99
    political ideology, 98, 100
    positivism, 97, 104
    praxis, 98, 99, 101–2, 105–6, 109
    reconstruction technique, 97–98, 103–7, 109
    social constructionism, 96–97, 104, 108–10
    social Darwinism, 98, 101
    social ideology, 98, 100
    social injustice, 96, 97
    utilitarianism, 97, 103–4
    value theory, 97, 103–4
  determinism, 117–19
  disciplinary status, 13
  dominant relations, 105, 106–7, 110–19
  economic ideology, 103–7, 108, 115
  emancipation, 114, 119
  evolutionary approach, 95, 98, 104–7, 114–19
  habits of thought, 109, 115–16
  immanent critique, 104–7, 114–15

Evolutionary Critical Theory *(continued)*
  leisure class, 104–5, 130–31, 133, 135–36
  ontological status, 35, 58, 62–63
  overview, 95–96, 160–62
  theory-practice gap, 96, 108–19
  totalitarianism, 115
  See also Good society; Power relationships; Public administration; Teleology
Evolutionary disciplinary matrix, 12–13
Evolutionary nature
  hermeneutics, 74–75
  institutionalism, 68–69
  See also Endogenous evolution
Evolutionary paradigm, 65, 78, 91, 92–94
Evolutionary theory, 12–13

F

Family resemblance approach
  critical theory, 79, 86, 96, 128
  disciplinary status, 18–25, 163
  public administration theory, 66–67, 77
Fictionalism, 170
Functionalism, 170, 171, 172

G

Good society
  conclusions, 157–59
  critical theory
    agency, 145–48
    coercive power, 149–50
    deconstruction technique, 142, 145
    democratic ideology, 141, 144
    discourse ethics, 149
    dominant relations, 141–44, 145–48, 149
    emancipation, 141–50
    emotional security, 144, 149

Good society
  critical theory *(continued)*
    empowerment, 143–44
    endogenous forces, 148
    ideal speech situation, 142, 143, 144
    identity, 143, 145–48, 149
    immanent critique, 142, 148
    individual choice, 140–50
    individual choice dilemma, 140–41, 144–48
    individual choice resolution, 141–42
    leisure class, 147, 153
    marginalization, 142, 143, 145–48
    minimalist state, 144–45
    negative dialectics, 145
    political/personal virtues, 141
    power relationships, 136, 137, 141–42, 144, 145–48, 149–50
    praxis, 141–42, 144–48
    reflective self-critique, 142–44, 148
    religiosity, 143, 146, 147
    shared interests, 149
    social constructionism, 142, 145
    social injustice, 142, 149
    theory-practice gap, 141–42
  Evolutionary Critical Theory
    agency, 154, 156
    critical theory implications, 154–57
    dominant relations, 151–54
    emancipation, 150–57
    endogenous forces, 154
    good society defined, 154–55, 156–57
    identity, 153, 156
    individual choice, 150–57
    language game, 154–55
    power relationships, 150–54, 156, 158–59
    social constructionism, 157
    theory-practice gap, 151

Good society *(continued)*
  ontological status, 56–57
  overview, 138–40
Good theory, 8–9, 35–36, 64

H

Habits of thought, 98, 100, 101–2, 109, 115–16, 133
Hermeneutics
  critical theory synergy, 77–78, 85–90
  institutional/behavioral/hermeneutic approach, 5, 34, 38–39, 64–65
  public administration theory, 64–65, 66, 73–77
Heuristic devices, 10–11, 14
Human agency. *See* Agency
Hyperliberalism, 58–59

I

Ideal form, 13–14
Ideal speech situation
  good society, 142, 143, 144
  power relationships, 121, 127
  theory-practice gap, 108–9, 110–11, 112, 113–14
Identity
  critical theory, 143, 145–48, 149
  Evolutionary Critical Theory, 153, 156, 162–64
  good society, 143, 145–48, 149, 153, 156
  public administration, 162–64
Identity crisis, 9, 12–13, 30
Identity thinking, 47–50
Immanent critique
  critical theory, 95, 96, 97, 99–100, 102–4, 108–9
    good society, 142, 148
    institutionalism synergy, 83–85, 90, 91–92
    power relationships, 126

Immanent critique *(continued)*
    Evolutionary Critical Theory, 104–7, 114–15
Individual beingness, 130–31
Individual choice
  critical theory, 140–50
  Evolutionary Critical Theory, 150–57
Institutionalism
  critical theory synergy, 77–78, 82–85, 94
  institutional/behavioral/hermeneutic approach, 5, 34, 38–39, 64–65
  ontological status, 37–38, 39, 57–58
  public administration theory, 64–65, 66, 68–73
Instrumentalism, 97, 112–14, 116, 122
Interpretivism, 171, 172
Intersubjective experience of good governance
  critical theory
    critical theory role, 77–78, 79
    evolutionary paradigm, 92–93
    public administration application, 90, 92
  disciplinary status
    clustering approach, 16
    cognitive approach subparadigm, 30–31
    collectivism, 29
    comparative methodology, 26, 27–28
    disciplinary matrix, 28, 31–32
    ethical approach subparadigm, 31–32
    orthodoxy utilization, 14, 25
    paradigms, 30–32
    political science framework, 27–28, 32
    politics-administration distinction, 25–26, 27–28
    popular sovereignty, 25–26
    positivism, 28
    power relationships, 28

Intersubjective experience of good governance
disciplinary status *(continued)*
  public administration defined, 25–32
  public administration origin, 25–32
  rationality, 27
  redefinition approach, 28–29
  research publications, 29–30
  scientific method, 26, 27, 31
  social science research, 26–28
  subparadigms, 30–32
  theoretical estrangement, 12
Evolutionary Critical Theory, 9, 121–22
good governance defined, 4–5
governance defined, 3–4
intersubjectivity defined, 4–5
ontological status
  agency, 51–52
  identity thinking, 48–50
  irreconcilable intersubjectivity, 50–51
  methodological implications, 5–6, 38–39
  power relationships, 58–59
  public administration core concerns, 57–58
  public administration subject matter, 35–36
  social justice, 60
  theoretical implications, 5–6, 37–38
public administration
  defined, 25–32
  epistemology, 172
  legitimacy, 166, 167, 168, 169
  origin, 25–32
public administration theory
  hermeneutics, 73–74, 75
  institutionalism, 69–70, 71
  objectives, 67
  social science research, 67–68

Intersubjective pluralism
  critical theory, 78
  ontological status, 50–51, 57–58
  public administration theory, 68
Intersubjectivity
  defined, 4–5
  disciplinary status, 10, 11, 19–22
  ontological status, 40–51
    identity thinking, 47–50
    intersubjective meanings/particulars, 44–47
    irreconcilable intersubjectivities, 50–51
  See also Intersubjective experience of good governance

L

Language game, 19–22, 77, 154–55
Latent functions, 105
Legal Foundations of Capitalism (Veblen), 101
Legitimacy, 165–69
Leisure class
  critical theory, 98–99, 100, 147, 153
  Evolutionary Critical Theory, 104–5, 130–31, 133, 135–36
Linguistic interconnections
  disciplinary status, 18–19
  genealogical connections, 19, 77
  public administration theory, 73, 76–77

M

Marginalization, 7, 142, 143, 145–48
Materialism, 52, 108
Melancholy science, 126, 166
Metaphysical approach, 103–4
Minimalist state, 121, 137, 144–45, 161
Mormons, 146
Mundane intersubjectivity, 41, 43, 46–47

N

Natural laws, 97, 104
Natural liberty, 61–62
Natural rights, 61–62, 100
Negative dialectics, 81–82, 90, 91–92, 98, 99, 145
Nihilism, 55–57
Normative theory
  critical theory, 83
  ontological status, 34, 40, 44–45, 55–57
  public administration theory, 64–65, 74–75

O

Ontological status
  agency
    determinism, 52–55
    intersubjective experience of good governance, 51–52
    materialism, 52
    methodological implications, 53
    power relationships, 53
    religiosity, 52
    socialization process, 52–53, 54–55
    sociological determinism, 52–53, 54–55
    theoretical implications, 53
    thesis limitations, 51–55
    utility maximization, 53
  conclusions, 62–63
  emancipation
    civil liberty, 61
    government regulation, 61
    identity thinking, 48–49
    natural liberty, 61–62
    natural rights, 61–62
    public administration core concerns, 58, 61–62
    socialization process, 61–62

Ontological status *(continued)*
  Evolutionary Critical Theory, 35, 58, 62–63
  identity thinking
    collectivism, 48–50
    communication, 47–48
    critical theory, 48
    emancipation, 48–49
    intersubjective experience of good governance, 48–50
    power relationships, 48–49
    relevance structures, 49–50
    religiosity, 48–49
    socialization process, 47–48
    thesis limitations, 47–50
  intersubjective experience of good governance
    agency, 51–52
    identity thinking, 48–50
    irreconcilable intersubjectivity, 50–51
    methodological implications, 5–6, 38–39
    power relationships, 58–59
    public administration core concerns, 57–58
    public administration subject matter, 35–36
    social justice, 60
    theoretical implications, 5–6, 37–38
  intersubjective meanings/particulars
    common subject matter, 44–47
    communication, 47
    empirical theory, 44–45
    mundane intersubjectivity, 46–47
    normative theory, 44–45
    relevance structures, 47
    socialization process, 46–47
    thesis limitations, 44–47
  intersubjectivity
    common subject matter, 40–44
    communication, 40–41
    empirical theory, 40

Ontological status
  intersubjectivity *(continued)*
    mundane intersubjectivity, 41, 43
    normative theory, 40
    presupposition suspension, 44
    socialization process, 41–43
    thesis limitations, 40–44
    transcendental intersubjectivity, 41–42, 43, 44
  irreconcilable intersubjectivity
    collectivism, 50–51
    democratic ideology, 50
    intersubjective experience of good governance, 50–51
    intersubjective pluralism, 50–51
    power relationships, 50–51
    socialization process, 50
    thesis limitations, 50–51
    zone of acceptance, 50–51
  methodological implications, 36
    action research, 39
    institutional/behavioral/hermeneutic approach, 38–39
    institutionalism, 39
    intersubjective experience of good governance, 5–6, 38–39
    qualitative research, 39
    situational analysis, 39
    subject matter comprehension, 38–39
    survey research, 39
    synergistic development, 5, 38–39
  nihilism
    empirical theory, 55–56
    good society, 56–57
    normative theory, 55–57
    positivism, 55–56
    religiosity, 56
    socialization process, 56–57
    thesis limitations, 55–57
  overview, 34–35
  power relationships
    agency, 53
    authoritarianism, 58–59

Ontological status
  power relationships *(continued)*
    cognitive authority, 58–59
    hyperliberalism, 58–59
    identity thinking, 48–49
    intersubjective experience of good governance, 58–59
    irreconcilable intersubjectivity, 50–51
    powerlessness, 59
    practical authority, 58–59
    public administration core concerns, 57, 58–59
    public spaces, 59
    socialization process, 59
    theory-practice gap, 58–59
  public administration core concerns, 57–62
    Evolutionary Critical Theory, 58
    institutionalism, 57–58
    intersubjective experience of good governance, 57–58
    intersubjective pluralism, 57–58
    zone of acceptance, 57–58
  public administration subject matter
    empirical theory, 34, 40
    good theory defined, 35–36
    intersubjective experience of good governance, 35–36
    normative theory, 34, 40
    positivism, 36
    scientific method, 34
    social reality, 34, 35–36
    social science research, 34
    theory defined, 35
  socialization process
    agency, 52–53, 54–55
    emancipation, 61–62
    identity thinking, 47–48
    intersubjective meanings/particulars, 46–47
    intersubjectivity, 41–43
    irreconcilable intersubjectivity, 50

Ontological status
socialization process *(continued)*
   nihilism, 56–57
   power relationships, 59
   social justice, 60–61
social justice
   cognitive authority, 59–60
   critique process, 60–61
   equality, 61
   injustice, 60
   intersubjective experience of good governance, 60
   priori standard, 60
   public administration core concerns, 58, 59–61
   socialization process, 60–61
synergistic development
   Evolutionary Critical Theory, 35
   institutional/behavioral/hermeneutic approach, 34
   methodological implications, 5, 38–39
   theoretical implications, 38
theoretical estrangement, 12–13
theoretical implications, 36–38
   institutionalism, 37–38
   intersubjective experience of good governance, 5–6, 37–38
   social science research, 36–37
   synergistic development, 38
   theory characteristics, 37–38
   transition points, 37
theoretical myth, 3, 34, 35
theoretical objectives, 34
thesis limitations, 39–57
See also Disciplinary status; Public administration theory
Orthodoxy utilization, 13–15, 21–23, 25, 28, 165

P

Pecuniary culture, 99, 100, 147
Political ideology, 98, 100
Political science framework, 27–28, 32, 163–64
Politics-administration distinction, 14, 25–26, 27–28
Popular sovereignty, 25–26
Positivism
   critical theory, 80–81, 84, 97, 104
   disciplinary status, 28
   ontological status, 36, 55–56
   public administration epistemology, 170, 171
   public administration theory, 68
Powerlessness, 59
Power relationships
   conclusions, 136–37
   critical theory
      administrative state, 122–29
      collectivism, 126
      conservatism, 128
      democratic ideology, 121, 127
      dominant relations, 122–23, 124–29
      emancipation defined, 121, 122, 125–26
      emancipation paradoxes, 126–30, 133–36
      empowerment, 128
      evolutionary paradigm, 94
      family resemblance approach, 128
      good society, 136, 137, 141–42, 144, 145–48, 149–50
      ideal speech situation, 121, 127
      immanent critique, 126
      institutionalism synergy, 82–85, 94
      instrumentalism, 122
      melancholy science, 126
      minimalist state, 121, 137
      power defined, 122–25
      power paradoxes, 126–30, 133–36
      reconstruction technique, 128
      structural determinism, 124
      utility maximization, 122–23
   disciplinary status, 20, 28

Power relationships *(continued)*
  Evolutionary Critical Theory
    administrative state, 129–36
    being power, 131–33, 136
    conservatism, 135–36
    dominant relations, 131–36
    drift, 132–36
    emancipation defined, 121–22, 130–36
    empowerment, 133
    endogenous evolution, 98–100, 102–3, 104–5, 121, 133–34
    good society, 150–54, 156, 158–59
    habits of thought, 133
    individual beingness, 130–31
    interstitial emancipation, 133
    intersubjective experience of good governance, 121–22
    leisure class, 130–31, 133, 135–36
    paradox resolution, 129–30, 133–36
    personal power, 130–31
    power defined, 130–36
    power theory, 129–36
    social constructionism, 121
    theory-practice gap, 110–11, 113–14, 115, 118–19, 129, 131
    truth regimes, 133–34
  ontological status, 48–49, 50–51, 53, 57, 58–59
  overview, 121–22
  public administration legitimacy, 166–69
  See also Dominant relations; Emancipation; Empowerment
Practical authority, 58–59
Praxis, 98, 99, 101–2, 105–6, 109
  critical theory/institutionalism synergy, 83, 85, 90, 91–92
  good society, 141–42, 144–48

Priori standard, 60
Public administration
  conclusions, 172–73
  core concerns, 57–62
  epistemology
    behaviorism, 171–72
    empiricism, 170–72
    fictionalism, 170
    functionalism, 170, 171, 172
    interpretivism, 171, 172
    intersubjective experience of good governance, 172
    paradigm synergy, 171–72
    positivism, 170, 171
    radical humanism, 171, 172
    radical structuralism, 171, 172
    realism, 170
    social constructionism, 170, 172
    theorists, 170
  Evolutionary Critical Theory, 160–62, 172–73
    epistemology, 170–72
    legitimacy, 165–69
  future of, 172–73
  identity
    disciplinary matrix, 162–64
    disciplinary status, 162
    ontological status, 162
    political science framework, 163–64
  legitimacy
    balance wheel, 169
    conservatorship, 169
    critical theory, 166
    democratic ideology, 165
    drift, 166, 169
    executive, 167–68
    intersubjective experience of good governance, 166, 167, 168, 169
    judiciary, 165–66
    legislature, 166–67
    orthodoxy utilization, 165
    power relationships, 166–69

Public administration *(continued)*
  origin
    family resemblance approach, 23–24
    intersubjective experience of good governance, 25–32
    overview, 160–62
    subject matter
      critical theory application, 90–91
      ontological status, 35–36, 40
    See also Critical theory; Disciplinary status; Ontological status; Power relationships
Public administration defined
  clustering approach, 16–18, 25
  family resemblance approach, 19–25, 29–30
  intersubjective experience of good governance, 25–32
  redefinition approach, 20–22, 28–30
Public Administration Review, 30
Public administration theory
  hermeneutics
    defined, 66
    empirical theory, 74–75
    empowerment, 66
    evolutionary nature, 74–75
    family resemblance approach, 77
    hermeneutic circle, 73–74
    intersubjective experience of good governance, 73–74
    linguistic interconnections, 76–77
    linguistic meaning, 73
    normative theory, 74–75
    objectives, 66, 73
    ontological hermeneutics, 75–76
    ontological status, 64–65
    power relationships, 76
    research limitations, 75–77
    research value, 73–75
    social constructionism, 74
    systems of thought, 76
  institutionalism
    agency, 68–69, 72–73

Public administration theory
  institutionalism *(continued)*
    defined, 66
    determinism, 72–73
    empowerment, 66
    evolutionary nature, 68–69
    holistic nature, 68–69
    intersubjective experience of good governance, 69–70, 71
    objectives, 66, 68–69
    objectivist approach, 70–71
    ontological status, 64–65
    research limitations, 70–73
    research value, 68–70
    socialization process, 69–70, 71, 72
    social reality, 69, 70, 72–73
    social value theory, 72
    unpredictable process, 69–70
    zone of acceptance, 71
  intersubjective experience of good governance
    hermeneutics, 73–74, 75
    institutionalism, 69–70, 71
    objectives, 67
    social science research, 67–68
  objectives
    clustering approach, 66–67
    collectivism, 66–67
    family resemblance approach, 66–67
    intersubjective experience of good governance, 67
  ontological status
    empirical theory, 64–65
    good theory, 64
    institutional/behavioral/hermeneutic approach, 64–65
    intersubjective meanings/particulars, 64
    normative theory, 64–65
    synergistic development, 64–65
  overview, 64–65

Public administration theory
(continued)
  research concerns
    authority, 67
    choice, 67
    decision making, 67
    legitimacy, 67
  social science research
    defined, 65
    empowerment, 66
    intersubjective experience of good governance, 67–68
    intersubjective pluralism, 68
    objectives, 65–66
    ontological status, 64–65
    positivism, 68
    research limitations, 67–68
    research value, 68
    social reality, 67
    transcendental intersubjectivity, 68
    zone of acceptance, 68
  See also Disciplinary status; Ontological status
Public spaces, 59

Q

Quakers, 147
Qualitative research, 39

R

Radical humanism, 171, 172
Radical structuralism, 171, 172
Rationality, 11, 27
Realism, 170
Reconstruction technique, 8–9, 65, 95, 97–98, 103–7, 109, 128
Reflective self-critique, 142–44, 148
Reflective synergy, 77–78, 80, 81
Relevance structures, 47, 49–50
Religiosity
  Christianity, 100, 108
  Evolutionary Critical Theory, 100, 108

Religiosity *(continued)*
  good society, 143, 146, 147
  Mormons, 146
  ontological status, 48–49, 52, 56
  Quakers, 147
  Reynolds v. *United States*, 146

S

Scientific method
  critical theory, 81–82, 90
  disciplinary status, 14, 26, 27, 31
  ontological status, 34
Situational theory
  disciplinary status, 10
  ontological status, 39
Social constructionism
  critical theory
    emancipation, 96–97, 108–10
    good society, 142, 145
    hermeneutics synergy, 86–90
    institutionalism synergy, 83–84
    public administration application, 91
    social science synergy, 81–82
  Evolutionary Critical Theory
    good society, 157
    power relationships, 121
  public administration
    critical theory application, 91
    epistemology, 170, 172
    theory, 74
Social Darwinism, 98, 101
Social ideology, 98, 100
Social injustice, 60, 96, 97, 142, 149
Socialization process, 69–70, 71, 72
  See also Ontological status
Social justice, 58, 59–61
Social reality
  critical theory, 79, 80, 88–90, 91
  ontological status, 34, 35–36
  public administration theory, 67, 69, 70, 72–73

Social science research
   critical theory synergy, 77–78, 80–82
   disciplinary status, 26–28
   ontological status, 34, 36–37
   public administration theory, 64–66, 67–68
Social science theory, 78
Social value theory, 72
Sociological determinism, 52–53, 54–55
Structuralism, 171, 172
Survey research, 39

T

Teleology
   critical theory reconstruction, 95, 98, 102–3, 150–52
   democratic teleological ideal, 7–8, 93
   Evolutionary Critical Theory, 102–3, 117–20, 121–22, 150–52, 165, 166, 167, 169, 171
   good society, 148–52
   hermeneutics, 76
   institutional/behavioral/hermeneutic synergy, 95
   institutionalism, 71
   intersubjective experience of good governance, 93
   knowledge/reality relationship, 171
   nihilism, 55–57
   nonteleological endogenous evolutionary approach, 95, 98, 121–22
   nonteleological evolutionary framework, 93–94, 95, 119–20, 150–54
   ontological status, 55–56
   power relationships, 121–22
   public administration
     epistemology, 171
     legitimacy, 165, 166, 167, 169
     theory, 71, 76
   religiosity, 56

Teleology *(continued)*
   social constructionism, 56
   systemic teleology, 102–3
   teleological boundaries, 171
   teleological crisis, 95–96
   teleological groundings, 7–8, 55–56, 76
   teleological moments, 7–8, 148–52
   theory-practice gap, 95–96, 117–19
   universalistic principles of behavior, 7–8
   universalistic principles of discourse, 7–8
   utilitarianism, 103–4
Theory of Business Enterprise, The (Veblen), 103, 105, 106–7, 152
Theory of the Leisure Class, The (Veblen), 98–99, 104–5, 130–31, 133, 147, 153
Theory-practice gap
   critical theory, 8, 93, 141–42
   disciplinary status, 11, 13
   Evolutionary Critical Theory, 96, 108–19
   emancipation, 114, 119
   good society, 151
   power relationships, 110–11, 113–14, 115, 118–19, 129, 131
   ontological status, 58–59
Theory-practice merge, 11
Totalitarianism, 115
Transcendental hermeneutics, 87–88, 90, 91–92
Transcendental intersubjectivity, 41–42, 43, 44, 68
Transition points, 4–5, 37, 90

U

Utilitarianism, 97, 103–4
Utility maximization, 53, 122–23

V

Value theory, 97, 103–4
Veblen, Thorstein
  critical theorist, 8–9, 98–103
  critical theory, 97–107
  evolutionary approach, 104–7, 114–19, 151–54
  Frankfort school, 95, 97, 108–9
  good society, 147, 150–54
  overview, 95–96
  publications, 98–99, 101, 103, 104–5, 106–7, 130–31, 133, 147, 152, 153

Veblen, Thorstein *(continued)*
  theory-practice gap, 108–19
  theory reconstruction, 97–98, 103–7

Y

Yoder v. *Wisconsin*, 147

Z

Zone of acceptance
  ontological status, 50–51, 57–58
  public administration theory, 68, 71

About the Authors

Charles Frederick Abel is an assistant professor of political science and public administration and director of the Center for Applied Social Research at Stephen F. Austin State University. Abel has a JD from Duquesne University, an MA from the University of Pittsburgh, and a PhD from the University of Maryland. He is the author of books and articles in the fields of public law, international political economy, and public administration theory. Abel also has an established research stream examining issues in higher education. In addition, supplementing his academic record, he has a distinguished career as a county attorney.

Arthur J. Sementelli is an assistant professor of public administration at Florida Atlantic University and a member of the editorial board of *Administrative Theory and Praxis*. He is the former undergraduate coordinator at Stephen F. Austin State University and assistant director of the Center for Applied Social Research. Sementelli has a PhD from Cleveland State University's Levin College of Urban affairs, an MPA from Gannon University, and an undergraduate degree from Carnegie Mellon. He is the author of articles on environmental policy, critical theory, and discourse theory.